Computational Intelligence Applications to Power Systems

T0140159

Computational Intelligence Applications to Power Systems

Yong-Hua Song   Allan Johns   Raj Aggarwal
School of Electronic and Electrical Engineering
University of Bath
England, UK

Science Press
New York/Beijing

Kluwer Academic Publishers
Dordrecht/Boston/London

Library of Congress Cataloging in Publication Data

ISBN 978-90-481-4711-3

Kluwer Academic Publishers incorporates
the publishing programmes of
D. Reidel, Martinas Nijhoff, Dr. W. Junk and MTP Press.

Sold and distributed in the U.S.A. and Canada
by Kluwer Academic Publishers,
101 Philip Drive, Norwell, MA 02061, U.S.A.

Sold and distributed in the People's Republic of China
by Science Press, Beijing.

In all other countries, sold and distributed
by Kluwer Academic Publishers Group,
P.O. Box 322, 3300 AH Dordrecht, The Netherlands.

PREFACE

Artificial intelligence is a multi-disciplinary subject which has attracted experts from psychology, computer science, engineering and social science. As we all agree, computers are wonderful machines which have very significantly changed the way we work and live. Computers can process large amounts of data quickly and provide an accuracy far beyond the capabilities of human beings. Today computers not only store, manipulate, and retrieve data, they are increasingly taking on the role of making important decisions. In the past two decades, the technology has moved from the data processing age, through the information manipulation era, and into our present world of simulating human problem solving capability; this has given rise to a new and exciting subject - Artificial Intelligence (AI).

Artificial Intelligence is defined as the study of the computations that make it possible to perceive, reason and act. From this definition, we can see that AI differs from most of psychology because of its greater emphasis on computation; also, AI differs from most of computer science because of the emphasis on perception, reasoning, and action.

From the perspective of goals, AI can be viewed as part engineering, part science:

■   The engineering goal of AI is to solve real-world problems using AI as a tool to simulate human problem solving capabilities. In its simplest form, the goal is to develop machines or computer programs that emulate human intelligence.

■   The scientific goal of AI is to better understand how humans solve problems. Building an intelligent machine requires an understanding of how humans develop, organise and use their intelligence.

Common sense tells us that humans have a large variety of intelligence. A human is able to learn from experience or hands-on or books, is able to handle uncertain situations, is able to adapt to the environment, is able to make a decision quickly, solely based on incomplete information, etc. Thus various AI techniques have been developed to achieve most of those important aspects of human intelligence. In this book, five major areas will be introduced, including:

■ Expert Systems: An Expert System is a computer program that emulates the decision making of a human expert. The emphasis of an expert system is the well-focused knowledge and inferring ability of a human expert. Due to historic reasons, knowledge-based systems dominate the early work in the AI area.

■ Fuzzy Logic: Fuzzy Logic is a tool to handle imprecise or ambiguous knowledge.

■ Artificial Neural Networks: Artificial Neural Network is a massively distributed processor motivated from the biological structure of a human brain that has a natural propensity for storing experimental knowledge and the ability to learn.

■ Genetic Algorithms: Genetic Algorithms are based on the principles of genetics and natural selection - Darwin's "survival of the fittest" strategy. They are adaptive search techniques.

■ Hybrid Systems: More recently, it is now becoming apparent that the integration of various intelligent techniques is a very important way forward in the next generation of intelligent systems.

Electric power systems have been around for many many years. Electrical engineers have developed various techniques to control, operate and manage the power networks efficiently. Then why do we need to introduce Artificial Intelligence techniques to Power Engineering? This is mainly due to the fact that there remains a large class of problems in power systems which cannot be solved by conventional numerical methods and are still largely solved by human experts either solely through experience or experience-based judgement in conjunction with the results obtained from the numerical based analysis and decision support systems. These problems are frequently characterized by the following features:

■ It is not always possible to develop a mathematical model for the problem which closely reflects the actual situation sufficiently.

■ The problem is such that some of the constraints are improperly specified and cannot be expressed mathematically.

■   The solution methodology employed by the human expert is not capable of expression in an algorithmic or a mathematical form.

■   The complexity and size of the problem is such that complete computational based solutions cannot be obtained in a timely fashion.

In addition, the high complexity and severity of power system operation makes operators require lengthy operational practice and periodical retraining to be fully qualified. Moreover, a rapid increase in the number of real-time messages has made the operator response more difficult.

Most of the material in this book has been developed and is being used for the undergraduate/Master's course at the University of Bath in the UK. The objective of the course is to teach some of the fundamentals of the major intelligent techniques and give an understanding as to how they have been applied to solving power system problems. This book should also be of great interest to Electrical Engineers or managers in the Electricity Supply Industry who want to understand and have an appreciation of the advances in this exciting subject.

ACKNOWLEDGEMENT

The authors gratefully acknowledge the IEEE and the IEE for their permission to use some of the materials in their publications. The assistance from Dr G S Wang in preparing the manuscript is appreciated. The authors express their special thanks to Mr P Lin of Science Press for his help in many aspects.

Dr Y H Song would also like to extend his sincere thanks to the many people who have given their support to his work in the area of Artificial Intelligence during the past several years, in particular, Professors Z M Xiong, P Y Wang, X X Zhou, Q Y Zhen, J D Gao, Y D Han, D Evans, and T E Dy-Liacco.

CONTENTS

CHAPTER 1
EXPERT SYSTEMS: AN INTRODUCTION

1.1 Major Characteristics of Expert Systems

Expert systems are computer programs that perform sophisticated tasks once thought possible only for human experts. The expert program is built from explicit pieces of knowledge extracted from human experts using artificial intelligence programming techniques such as symbolic representation, inference, and heuristic searches. Knowledge-based systems can be distinguished from other branches of artificial intelligence by their emphasis on domain-specific knowledge, rather than more general problem-solving strategies. Because their strength derives from such domain-specific knowledge rather than more general problem-solving strategies, expert systems are often called "knowledge-based".

1.1.1 Expert system structure

Expert systems typically have the four components illustrated in Figure 1.1.

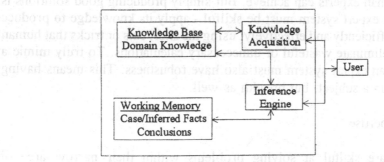

Figure 1.1 General structure of an expert system

(i) The knowledge base consists of highly specialized knowledge on the problem area as provided by the expert.

(ii) The inference engine is the knowledge processor which is modeled after the expert's reasoning. The engine works with available information on a given problem, coupled with the knowledge stored in the knowledge base, to draw conclusions or recommendations.

(iii) The knowledge-acquisition interface assists experts in expressing knowledge in a form suitable for incorporation into a knowledge base.

(iv) The user interface assists users in "consulting" the expert system, prompting them for information required to solve their problem, displaying the program's conclusions, and explaining its reasoning. Generally, these interfaces attempt to provide the user with most of the capabilities they would have if they were interacting with a human expert.

1.1.2 Characteristics of an expert system

(i) Possesses expert knowledge

An important feature of the knowledge used in an expert system is that it embodies the expertise of a human expert. It is this expertise that we try to capture and encode in an expert system. It includes both domain knowledge and problem-solving skills.

Expertise is a resource held by a few individuals who can successfully apply it to solve problems that are out of reach of others. An expert system must perform well, that is, it must achieve the same levels of performance in the domain of interest that human experts can achieve. But simply producing good solutions is not enough. An expert system must be skilful - apply its knowledge to produce solutions both efficiently and effectively, using the short cuts or tricks that human experts use to eliminate wasteful or unnecessary calculations. To truly mimic a human expert, an expert system must also have robustness. This means having not only depth in a subject, but breadth as well.

(ii) Focuses expertise

Most experts are skilful at solving problems within their narrow area of expertise. Thus an expert system has depth, i.e., it operates effectively in a narrow domain containing difficult, challenging problems.

(iiii) Reasons with symbols

When human experts solve problems, particularly the type we consider appropriate for expert systems work, they do not do it by solving sets of equations or performing other laborious mathematical computations. Instead they choose symbols to represent the problem concepts and apply various strategies and heuristics to manipulate these concepts. An expert system also represents knowledge symbolically, as sets of symbols that stand for problem concepts.

(iv) Reasons heuristically

Experts are adept at drawing on experience to help them efficiently solve current problems. Through their experience, they form a practical understanding of the problem and retain it in the form of rules-of-thumb or heuristics.

(v) Permits inexact reasoning

Expert systems have demonstrated considerable success in applications that require inexact reasoning. Such applications are characterized by information that is uncertain, ambiguous or unavailable, and by domain knowledge that is inherently inexact.

(vi) Explains results

Most current expert systems have what is called an explanation facility. This is an ability to explain how the system arrived at its answers. Most of these explanations involve displaying the inference chains and explaining the rationale behind each rule used in the chain. The ability to examine their reasoning processes and explain their operations is one of the most innovative and important qualities of expert systems.

(vii) Makes mistakes

No matter how proficient some experts are, they share one shortcoming with the rest of us; they are only human and can make mistakes. We recognize this possibility any time we consult with an expert. Since your task is to capture as closely as possible the knowledge of the expert, you should recognize that your system, like its human counterpart, can make mistakes.

1.1.3 Pro and cons of expert systems

(i) Expert systems versus conventional programs

Conventional programs address problems where the information is complete and exact. However, if data are faulty or missing, a conventional program may provide no results. On the other hand, the types of problems that expert systems work on are less structured than conventional programs because the information available may not be sufficient to arrive at an exact solution. However, an expert system may still arrive at reasonable conclusions - even if they are suboptimal. Some differences between expert systems and conventional programs are shown in Table 1.1.

Table 1.1 Expert systems versus conventional programs

Conventional Programs	Expert Systems
Numeric	Symbolic
Algorithmic	Heuristic
Information and control integrated	Knowledge separation from control
Difficult to modify	Easy to modify
Precise information	Uncertain information
Repetitive process	Inferential process
Final result given	Recommendation with explanation
Optimal solution	Acceptable solution
Effective manipulation of large data bases	Effective manipulation of large knowledge bases

(ii) Strengths and weaknesses of expert systems

There are some excellent reasons for using artificial expertise to augment human reasoning. Some of these advantages are summarized in Table 1.2. Due to these advantages, an expert system can be built to replace a human expert or assist a human expert in solving real-world problems.

Table 1.2 Comparison of human expert and an expert system: the good news

Human expert	Expert system
Perishable	Permanent
Difficult to transfer	Easy to transfer
Difficult to document	Easy to document
Unpredictable	Consistent
Expensive	affordable

When applying expert systems to a particular application, their weaknesses must be borne in mind as well. Some of these weaknesses are listed in Table 1.3.

Table 1.3 Comparison of human expert and an expert system: the bad news

Human expert	Expert system
Creative	Uninspired
Adaptive	Needs to be told
Sensory experience	Symbolic input
Broad focus	Narrow focus
Commonsense knowledge	Technical knowledge

1.2 Rule-based Expert Systems

1.2.1 Knowledge representation

Knowledge is power. This short phrase is often used to emphasize the importance of knowledge to an expert system. It has long been recognized that the performance of an expert system is directly related to the quality of knowledge the system has on a given problem. Knowledge is an abstract term that attempts to capture an individual's understanding of a given subject. However, when building an expert system, we do not attempt to capture all of the expert's knowledge. Rather, we target and focus on the expert's knowledge of a well-focused topic from the subject area. This is called domain-specific knowledge. After knowledge from the expert on some well-focused domain has been acquired, a way to encode it in the expert system is needed. This is known as knowledge representation. There are various types of knowledge human experts use, such as: procedural knowledge, declarative knowledge, meta-knowledge, heuristic knowledge and structural knowledge. Through the efforts of researchers in artificial intelligence, a number of effective ways of representing knowledge in a computer have been developed, such as rules, sematic and frame. In this section, the commonly used technique -rule - is introduced.

In expert system jargon the term rule has a much narrower meaning than it does in ordinary language. It refers to the most popular type of knowledge representation technique, thé rule-based representation. Rules provide a formal way of representing recommendations, directives, or strategies; they are often appropriate when the domain knowledge results from empirical associations developed through years of experience solving problems in an area.

The rule's structure logically connects one or more premises contained in the IF part, to one or more conclusions contained in the THEN part. For example:

> IF   the current difference is larger than zero
> THEN a fault occurs

In general, a rule can have multiple premises joined with AND statements, OR statements, or a combination of both. Its conclusion can contain a single statement or a combination joined with an AND. The rule can also contain an ELSE statement, that is inferred to be TRUE if one or more of its premises are false.

1.2.2 Inference techniques

Humans solve problems by combining facts with knowledge. They take facts about a specific problem and use them with their general understanding of the problem domain to derive logical conclusions. This process is called reasoning. Expert systems model the reasoning process of humans using a technique called inference. An expert system performs inference using a module called the inference engine. It is a process which works with current information to derive further conclusions. It combines facts contained in the data base with knowledge contained in the knowledge base.

1.2.2.1 Forward-chaining

The solution process for some problems naturally begins by collecting information. This information is then reasoned to infer logical conclusions. This style of reasoning is modelled in an expert system using a data-drive search; it is also called forward-chaining. Forward-chaining is a inference strategy that begins with a set of known facts, derives new facts using rules whose premises match the known facts, and continues this process until a goal state is reached or until no further rules have premises that match the known or derived facts.

The simplest application of forward-chaining in a rule-based expert system proceeds as follows. The system first obtains problem information from the user and places it in the data base. The inference engine then scans the rules in some predefined sequence looking for one whose premises match the contents in the data base. If it finds a rule, it adds the rule's conclusion to the data base and then cycles and checks the rules again looking for new matches. On the new cycle, rules that previously fired are ignored. This process continues until no matches are found. At this point the data base contains information supplied by the user and inferred by the system. Figure 1.2 shows a flowchart of this simple form of forward-chaining.

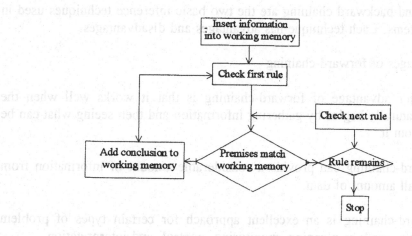

Figure 1.2 Forward-chaining inference

1.2.2.2 Backward-chaining

Forward-chaining is a good inference technique if we are working with a problem that requires us to begin with information and then derive logical conclusions. In other problems, we begin with a hypothesis and then attempt to prove it by gathering supporting information. This style of reasoning is modelled in an expert system using a goal-driven search; it is also called backward-chaining.

A backward-chaining system begins with a goal to prove. It first checks the data base to see if the goal has been previously added. This step is necessary since another knowledge base may have already proven the goal. If the goal has not previously been proven, the system searches its rules looking for one (or more) that contains the goal in its THEN part. This type of rule is called a goal rule. The system then checks to see if the goal rule's premises are listed in the data base. Premises not listed then become new goals (also called subgoals) to prove that may be supported by other rules. This process continues in this recursive manner, until the system finds a premise that is not supported by any rule. When a primitive is found, the system asks the user for information about it. The system then uses this information to help prove both the subgoals and the original goal.

1.2.2.3 Combining forward- and backward-chaining

Forward and backward chaining are the two basic inference techniques used in expert systems. Each technique has advantages and disadvantages.

(i) Advantages of forward-chaining

(1) A major advantage of forward-chaining is that it works well when the problem naturally begins by gathering information and then seeing what can be inferred from it.

(2) Forward-chaining can provide a considerable amount of information from only a small amount of data.

(3) Forward-chaining is an excellent approach for certain types of problem solving tasks, such as planning, monitoring, control, and interpretation.

(ii) Disadvantages of forward-chaining

(1) One of the major disadvantages of a forward chaining system is that it may have no means of recognizing that some evidence might be more important than others. The system will ask all possible questions, even though it may only need to ask a few questions to arrive at a conclusion.

(2) The system may also ask unrelated questions.

(iii) Advantages of backward-chaining

(1) One of the major advantages of a backward-chaining system is that it works well when the problem naturally begins by informing a hypothesis and then seeing if it can be proven.

(2) Backward-chaining remains focused on a given goal. This produces a series of questions on related topics, a situation that is comfortable for the user.

(3) Whereas a forward-chaining system attempts to infer everything possible from available information, a backward-chaining system searches only that part of the knowledge base that is relevant to the current problem.

(4) Backward-chaining is an excellent approach for certain types of problem solving tasks, such as diagnostics, prescription, and debugging.

(iv) Disadvantages of backward-chaining

The principal disadvantage of a backward-chaining system is that it will continue to follow a given line of reasoning even if it were to drop it and switch to a different one.

Many expert systems use both forward and backward-chaining. This is typically seen in applications where different tasks are naturally performed in either a data-driven or goal-driven fashion. There are two methods to combine them. The first method relies on separate systems, each with its own inference strategy. The second method incorporates demons within a backward-chaining system that act in a forward-chaining fashion. A demon rule is a rule that fires whenever its premises match the contents of the data base.

1.2.3 Pro and cons of rule based system

Rule-based systems are currently the most popular choice of knowledge engineers for building expert systems. This popularity has grown out of the large number of successful rule-based systems built from the abundance of available rule-based expert system development software.

(i) Advantages of rule-based systems

(1) Natural expression

For many problems, humans naturally express their problem solving knowledge in IF...THEN type statements. The ease of capturing this knowledge in a rule makes a rule-based approach an attractive choice for the design of the expert systems.

(2) Separation of control from knowledge

This feature permits you to change the system's knowledge or control separately. Thus additional rules can easily be added allowing for a graceful expansion of the system's knowledge.

(3) Utilization of heuristic and uncertain knowledge

A typical trait of human experts is that they are particularly adept at using "rules of thumb" or "heuristics" to help them to solve a problem efficiently. Rule based systems are well suited for working with these heuristics. Rules can easily be written to capture uncertain knowledge.

(ii) Disadvantages of rule-based systems

(1) Require exact matching

The rule-based system attempts to match the antecedents of the available rules with the facts in the working memory. For this process to be effective, this match must be exact, which in turn requires a strict adherence to consistent coding.

(2) Can be slow

Systems with a large set of rules can be slow.

1.2.4 Outside rule based system

Although rule-based expert systems are currently the most widely used, during the 1990s the movement will be towards object-oriented design. Object-oriented programming is a relatively new method for designing and implementing expert systems. Object-oriented programming centres around several major concepts: abstract data types and classes, encapsulation, inheritance, and polymorphism. The architecture of an object-oriented software system is built around a set of classes that characterize the behaviour of all the underlying data in the system. Objects from each class are manipulated by invoking the methods of the class, that is, sending messages to the objects. These messages represent the actions that are taken on the set of objects. Object-oriented programming focuses on the data to be manipulated rather than on the procedures that do the programming. The combination of object-oriented programming with frame based knowledge representation can create a powerful tool for addressing complex problems.

1.3 Applications in Power Systems

1.3.1 Building an expert system

1.3.1.1 Stages of expert system development

Expert system development can be viewed as essentially six interdependent and overlapping phases: assessment, knowledge acquisition, design, test, documentation and maintenance.

An expert system gains its power from the knowledge it contains; it is therefore very important that every effort be made to ensure that the knowledge that goes into the system effectively captures the domain expertise. In fact, knowledge acquisition has long been recognized as a bottleneck in the process of development of knowledge-based systems and remains the bigger challenge. The common knowledge sources are human experts, end-users and literature. In the early days, knowledge acquisition was done manually through a knowledge engineer interacting with a domain expert. This process was long, tedious and unsatisfactory. Domain experts are often not conscious of the basis of their decisions and therefore have difficulty articulating them in an explicit form. These features have given a strong impetus to the development of techniques for automated knowledge acquisition. The variety of techniques developed over the past few years include: (i) decision tree methods; (ii) progressive rule generation; (iii) explanation-based learning; (iv) methods that extract knowledge from neural networks and (v) genetic algorithm approaches. These all well documented in reference [7].

1.3.1.2 Expert system environment (tools)

Expert system tools are programming systems that simplify the job of constructing an expert system. They range from very high-level programming languages to low-level support facilities. Figure 1.3 illustrates the types of tools available for expert system building.

Compared with conventional high-level languages such as BASIC and FORTRAN, the features of most AI languages are: (1) symbol based and (2) descriptive. They can be classified as: (1) LISP, (2) Prolog, (3) object-oriented languages and (4) expert system shells. As an example, Prolog will be briefly introduced.

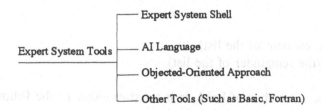

Figure 1.3 Expert system tools

(i) Features of Prolog

* Prolog is a declarative language. Facts and relationships about the system under investigation are encoded by making this information part of the knowledge base of the system. The user can then query the system for information either explicitly stated in the knowledge base or which is implied from this information.

* Prolog uses the language of predicate calculus. In predicate calculus, objects are represented by terms which may have one of the following forms:

A constant symbol representing a concept or single individual. A constant symbol is equivalent to a Prolog atom, and examples include greek, alberta, and peace (note lower case convention for constants).

A variable symbol which may represent different individuals at different times. Variables must be introduced along with a quantifier. Examples of Prolog variables are X and Roman (note upper case convention for variables).

A compound term which consists of a function symbol in conjunction with an ordered set of terms as its arguments. The function symbol represents how the function depends on the arguments. Thus, the prolog syntax for the relationship (predicate) "Steven likes Apple" takes the form: likes(steven,apple). After building this data base,a query of the form: likes(steven, apple)? will generate the response **yes.

* Prolog handles lists and recursion naturally. A very useful, recursively defined predicate is member(A,L)?which answers the question, "Is A member of list L?" If we have a list defined as:

L=[1,b,c,d,...]
then we may write
L=[Head|Tail}
where
Head=a (the first element of the list)
Tail=[b,c,d,...] (the remainder of the list)

A is defined to be a member of list L is it satisfies either of the following two rules:

   A is a member of list L if A is the first element of L.
   If A is not the first element of L, then A is a member of L
   if and only if A is a member of the tail of the list L.

These two rules may be encoded recursively in Prolog by the statements:

member(A,[A|_]).
member(A,[_|Tail]) :-member(A, Tail).

Note that the "_" symbol signifies "this variable matches anything".
* Prolog provides for every efficient coding for problems requiring inference.
That is, to solve the logical inference problem:

All humans are mortal.
Socrates is a human.

Is Socrates mortal?

can easily be encoded in Prolog as

mortal(X) :- human(X).
human(socrates).

mortal(socrates)?

to which Prolog responds:

**yes

(ii) Syntax of Prolog

The general structure of Prolog programs can be summarized by the following
four statement types:
* Facts which are represented in the Prolog data base by predicate clauses of the
form:

like(john, mary).

This is the fact that John likes Mary.

* Rules

A Prolog rule is a general statement about objects and their relationships. It is
represented by the general statement form:

X :- Y.

which may be interpreted in English as X is true if Y is true. X is defined as the head of the rule and Y is the body.

* Questions:

Once the data base is established by entering the appropriate facts and rules, the user queries the system by the use of Prolog questions. For instance, assume that our data consisted of one fact and one rule:

professor (mr_jones).
poor (Person) :-professor(Person).

Then we ask

poor(mr_jones)?

and Prolog responds:

**yes

* Commands:

These are imperative denoted by the punctuation key exclamation point which tell Prolog to take some action. Example of Prolog-86 commands include:

dir ('a:')!   %directory of drive a:
load('a:mwf1.pro')!   %load a file from a:

Note that "%" symbol denotes a comment in Prolog.

* Arithmetic in Prolog-86 can work with integers in the range from -32.268 to 32.767 and provides a number of arithmetic functions such as:

X is 35 + 40? % Addition
X=75

X is 35/40?   %Integer divide
X=0

* Logical connectives

A number of logical connectives are available for comparing atoms and numbers:

75=75?   % Equality test
**yes
45> =35?   % Greater than, equal
**yes

* The "cut" in Prolog

Prolog performs its amazing feats of inference by the backtracking process - satisfying a goal or causing a goal to fail. It is easy to show that there are many instances when this automatic backtracking may lead to unpleasant or disastrous results. To handle these cases, the special operator "cut" is introduced. The symbol for cut is "!"

1.3.2 Overview of expert systems in power systems

The application of knowledge-based systems to power systems is a relatively new research area: the first applications were in the security assessment and system restoration areas proposed in 1981. Since then, many more have been reported. Good general surveys of the subject can be found in references 4, 5, and 6. From the literature search, the ten most commonly investigated expert systems areas are alarm reduction, fault diagnosis, steady-state and dynamic security assessment, restoration, remedial controls, environments for operational aids, substation monitoring and control, and maintenance scheduling. Figure 1.4 reproduced from ref [4] gives an overview of expert systems in power systems

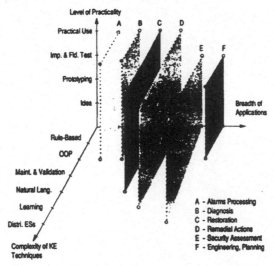

Figure 1.4 Overview of expert systems in power systems

References

[1] J Durkin, Expert Systems: Design and Development, Macmillan Publishing Company, 1994

[2] D A Waterman, A Guide to Expert Systems, Addison-Wesley, Reading, Mass., 1986

[3] A Barr, E A Feigenbaum, The Handbook of Artificial Intelligence, William Kaufman, 1981

[4] C C Liu, Eds., Special Issue on Knowledge-Based Systems in Electric Power Systems, Proceedings of the IEEE, Vol.80, No.5, 1992

[5] T S Dillon, M A Laughton, Eds., Expert System Applications to Power Systems, UK, Prentice Hall international, 1990

[6] M Huneault, C Rosu, R Manoliu, F D Galiana, A Study of Knowledge Engineering Tools in Power Engineering Applications, IEEE Trans on Power Systems, Vol.9, No.4, 1994

[7] S Sestito, T S Dillon, Automated Knowledge Acquisition, Prentice Hall, 1994

CHAPTER 2
INTELLIGENT ALARM PROCESSOR

2.1 Alarm Processing in EMS

Alarms are generated for a variety of system conditions: Circuit breaker changes; current limit exceeded; frequency deviation; voltage deviation; operation of protective equipment; nonfunctioning of remote controls; area control quantities out of limit, etc.

Modern EMS systems have as many as 5000 measurements and 16000 indicating devices being processed. Such an avalanche of messages can hamper operators in the process of making a quick response to disturbances. Modern energy management systems all have some form of alarm processing to alert the operators to power system parameters that are out of normal range or to changes that may affect the operation of the power system. An alarm processor is thus used to improve both the form and the content of the messages presented to the operator. The primary goal of these processors is to reduce the amount of information which the operator must process. The importance of each piece of information should be evaluated before it is presented to the operator and only the truly important alarms should be displayed. This threshold of importance can be reached by combining primitive alarm messages. These processors should also focus the attention of the operator and help him or her to track the evolution of the state of the power system by providing a summary of the abnormal conditions. Alarm processors are usually designed to handle all possible types of alarms with intent to present a clear picture of the global situation to the operator. Shorter response time is often required. A related topic is fault diagnosis which will be discussed in chapter 3. Fault diagnosis systems analyze only the alarms which are needed to locate a fault with the intention of finding a precise justification for a set of symptoms.

Several methods for effective alarm processing have been suggested and in some cases implemented. Filtering mechanisms, priority and grouping schemes as well as message routing procedures have been inserted into conventional alarm processing programs. While these approaches do reduce the number of alarms and raise the awareness of some of the most important ones, they cannot be easily used to synthesize messages with a higher information content. The use of combinatorial processing to reduce the number of alarms has also been proposed but with a lack of flexibility.

Processing the alarms with a rule-based expert system offers considerably more flexibility than any of the other approaches. A rule based system can be used to filter and prioritize alarms. Alarms can be combined and correlated with information gathered from other sources. The mental process of an operator trying to isolate the origin of a disturbance and select a course of action can also be emulated. Finally, the separation between the procedural knowledge and the maintenance knowledge greatly facilitates the development and the maintenance of these systems.

2.2 Basic Structure of Alarm Processing Expert System

Figure 2.1 illustrates the basic structure of an alarm processing expert system. There are two blocks which are the basis of any expert system i.e. the knowledge base and inference engine. The system introduced in this section is based on that described in reference[2].

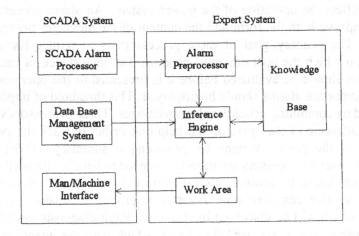

Figure 2.1 The basic structure of an alarm processing expert system

2.2.1 The rule base

The collection of knowledge in the IAP is referred to as the knowledge base. One way of organizing the knowledge is to form rules and facts. The rules contain accumulated knowledge in the form of IF-THEN constructs. The facts in the knowledge base are collected pieces of information related to the problem at hand such as power system elements and topology, the alarm message arriving at the control centre and the expert system conclusions. Rules express the relationship between facts.

An example of the use of production rules is as follows:

IF

       Alarm type is "AL" or "AL1" or "AL2", and this is a bus alarm, and no KV alarm is issued for nearby bus

THEN Print alarm

The rule base has two levels of rule: (i) metalevel control rules and (ii) object-level rules. The metalevel control rules contain heuristics that guide the inference engine in forming hypotheses. Heuristics are the rules of thumb that an expert system uses when solving a problem. The heuristics are used to narrow the search space for a solution. The backward chaining object-level rules are generic in nature and not related to any particular station. Portability of the rule base is accomplished by incorporating general principles in the goal driven object-level rules with station specific information in the forward chaining object-level rules and in the configuration nodes.

Figure 2.2 shows how the metalevel control rules are internally organized. Each metalevel node is the head of a linked list of one or more premise nodes. A premise node contains a premise clause. All the premise clauses of a rule must be true in order for the rule to fire. The inference engine scans the metalevel control list beginning with "metahead", the head node. If all of the premise clauses of a rule are true, the inference engine fires the metalevel control rule triggering one of the two metalevel actions, context switching or goal selection. In the case of goal selection, the metalevel node contains the goal, or hypothesis, that the object-level rules attempt to prove. The station node shown in Figure 2.2 has two working hypotheses. Each hypothesis is represented by a hypothesis node linked to the station node. A hypothesis node points to the metalevel node that it is associated with. When a working hypothesis is proved to be valid by the inference engine, the hypothesis node is linked to the station nodes' validated conclusion chain.

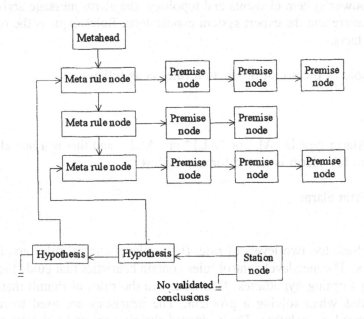

Figure 2.2 The internal organization of the metalevel control rule node

The object-level rules used for backward chaining are organized as shown in Figure 2.3. These goal driven rules are general rules that apply to any station. The unique characteristics of a station are organized as forward chaining object level rules and are not shown in the figure. Each general object-level rule is the head node of a linked list of one or more premise nodes. There is a separate premise node for each clause in the general rule premise. During backward chaining, the inference engine uses the general rule conclusions as goals. To facilitate the search for a general object-level rule, the conclusions are inserted into a hash table. A hashing function computes the index into the table based on the conclusion name. An overview chain is created if two or more conclusions hash to the same slot in the table. Each general rule conclusion node points to its associated rule node. The hashing scheme greatly reduces the time needed to search for a general rule conclusion during backward chaining.

An example is shown to describe the operation of the IAP for processing low bus voltage. In this example, the bus has three bus section. The following shows some metalevel control and object-level rules. Object-level rule names are enclosed with colons and metalevel control rules are surrounded with double colons. Comments are enclosed with /* and */.

(i) Metalevel rule examples

/* For the metalevel control rule called "MetaRule 1",
   establish conclusion "MC1" if the point that changed states
   is on a bus. */

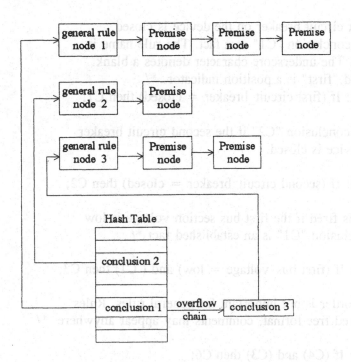

Figure 2.3 The general rule node layout

::MetaRule_1:: If (device = bus) then MC1;

/* If metaRule 1 has fired,determine the number of bus sections
   and the number of circuit breakers on the bus. From hypothesis
   "low bus voltage" with general rule conclusion "C6" as the goal
   if the directive results match the values. */
   ::MetaRule_3:: If (MC1) and (@count_bus_sections=3) and
           (@count_bus_breakers=2) then MC3 SC6 low_bus_voltage;

/*If the changes of state is a high amperage alarm, form an

insufficient supply to a zone hypothesis with general rule
conclusion "R15" as the goal */

::MetaRule_20:: If (point = line_current) then MC20
                    SR15 insufficient_trans.-capability_to_a_zone;

(ii) General rule examples

/* If the first circuit breaker on the device is closed,
   establish conclusion "C1" as a fact. The rule name is
   "Rule 1". The underscore character denotes a blank.
   The word "first" is a position indicator. */
   :Rule_1: If (first circuit_breaker = closed) then C1;

/* Establish conclusion "C2" if the second circuit breaker
   on the device is closed.   */

   :Rule_2: If (second circuit_breaker = closed) then C2;

/* The rule is fired if the first bus section voltage is low
   and conclusion "C1" is an established fact */

   :Rule_3: If (first bus_voltage = low) and (C1) then C3;

/* The rule order is not important for general rules. Rules
   are entered free format. comments may appear anywhere. */

   :Rule_6: If (C4) and (C3) then C6;

/* Fire this rule if the third bus section voltage is low and
   conclusions "C2" and "C5" are established facts. */

   :Rule_4: If (third bus_voltage = low) and (C1) and (C2) then C4;

/* If the second bus section voltage is low, establish
   conclusion "C5". */

   :Rule_5: If (second bus_voltage = low) then C5;

/* Fire this rule if the bus voltage on the bus at the opposite
   end of transmission line is low and the bus voltage in the

current station is normal. */

:Rule_15: If (opposite bus_voltage = low) and
                   (adjacent bus_voltage = Normal) the R15;

2.2.2 The inference engine

The inference engine is a part of an expert system which derives the logical conclusion, or conclusions, that can be obtained from the available data. The inference engine uses a knowledge base made up of production rules which describe in detail what logical conclusions can be drawn from each piece of data.

One of the questions facing all developers of real time expert systems is what kind of expert system to use. The most used inference techniques are forward, backward and mixed chaining. The difference between forward and backward chaining is in the method by which the facts and rules are searched. In forward chaining, the inference engine searches the IF portion of the rules. When a rule is found in which the entire IF portion is true, the rule is fired. Forward chaining is a data driven approach because the firing of rules depends on the current data. In backward chaining, the inference engine begins with a goal that is to be proved. It searches the THEN portion of the rules looking for a match. When a match is found, the IF portion of the rule is established. The IF portion may consist of one or more unproven facts and these facts become separate goals that must be proved. Backward chaining is goal driven because the order of the firing of the rules is done in an attempt to prove a goal. In mixed chaining, when the inference engine starts execution, it first executes a global rule set; all other rule sets are executed by firing a rule which has a looping statement containing the rule set name. When the inference engine executes any rule set it first looks for parameters called "find first" parameters. These are parameters which are designed to be set to a value by calling an auxiliary procedure. When all the find first parameters have been given a value, every rule that uses at least one find first parameter is executed. Further forward chaining may occur at this point, if the logic allows.

Metalevel control rules improve system performance by selecting the object-level rules. The object-level rules in the IAP are the forward and backward chaining rules. Metalevel actions provide context switching and goal selection as shown in Figure 2.4. When a SCADA alarm arrives, the metalevel control rules are used by the inference engine to generate one of the two metalevel actions. A context switching action selects data driven, or forward chaining rules. That is, the metalevel control rules are used to elect the proper context based on the incoming alarm. When the context is selected, metalevel control rules are used to produce one or more goal selection actions. A goal selection action selects a hypothesis that the goal driven, or backward chaining rules, attempt to prove. As can be seen, object-level actions result from applying the forward and backward chaining object-level rules.

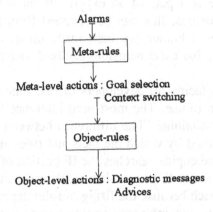

Figure 2.4 Metalevel control

The inference engine does not operate in an algorithmic, or step-by-step manner. Rather it is data driven. The selection of the next rule to fire depends on current hypotheses and facts related to a station. The inference engine continues to fire rules until there are no more rules to fire. It then suspends processing the hypothesis until another change of state occurs at which point the inference engine continues to prove the unsolved goal. By organizing information on a station basis, the inference engine can continue from its suspended state when processing alarms for the station.

2.3 Example Implementations

The prototype system described in section 2.2 was implemented in Wisconsin Electric of the USA. Some of the intelligent alarm processors described in the technical literature are listed in Table 2.1.

Table 2.1 Some intelligent alarm processors described in the technical literature

Utility	Location	Status
Consolidated Edison Company	New York, NY, USA	Operational Since June 1989
Northern States Power Company	Minneapolis, MN,USA	Operational Since October 1989
Electricite de France	Pairs, France	Prototype
Wisconsin Electric Power Company	Milwaukee, WI, USA	Prototype
Chubu Electric Power Company	Nagoya,Japan	Field testing
Hydro-Quebec	Montreal, PQ, Canada	Field testing
Hydro-Quebec	Montreal, PQ, Canada	Field testing
Energie-Versorging Schwaben,AG	Wendlingen, Germany	Field testing
Public Utilities Board of Singapore	Singapore	Operational Since November 1990
Vienna Electricity Board	Vienna, Austria	Prototype

Reference

1. B F Wollenberg, Feasibility study for an energy management system intelligent alarm processor, IEEE Trans Power Systems, Vol. PWRS-1, No.2, 1986, pp.241- 247
2. D B Tesch, D C Yu, L M Fu, K Vairavan, A Knowledge-based alarm processor for an energy management system, IEEE Trans Power Systems, Vol.5, No.1, 1990, pp.268 - 275
3. D S Kirschen, B F Wollenberg, Intelligent alarm processing in power systems, Procs of the IEEE, Vol.80, No.5, 1992, pp.663 - 672
4. R W Bijoch, S H Harris, T L Volkman, J J Bann, B F Wollenberg, Development and implementation of the NSP intelligent alarm processor, IEEE Trans Power Systems, Vol.6, No.2, 1991, pp.806 - 812
5. CIGRE Working Group report, T S Dillon, Survey on expert systems in alarm handling, Electra No.139, 1991, pp.133 - 151

CHAPTER 3
INTELLIGENT FAULT DIAGNOSIS
IN POWER SYSTEMS

3.1 Fault Diagnosis in Power Systems

The ultimate purpose of a power system is to transport electrical power from the power generation station to the consumer. As a result, it is configured from transmission and substation facilities distributed over a large geographical area. In order to achieve stable supply of electrical power, the power system must be extremely reliable. It is inevitable, however, that accidents such as lightning strokes, collisions with transmission lines, and failures due to ageing of equipment and random failures will occur. When a fault due to these causes does occur, it is imperative to limit the impact of outages to the minimum and to restore the faulted facilities as quickly as possible. This requires that the location and nature of the fault first be identified. This identification function is referred to as "fault diagnosis of power systems." This fault diagnosis function is then the most basic fault handling function of power system supervisory and control systems such as energy management systems (EMS) and supervisory control and data acquisition (SCADA) systems.

Fault diagnosis can be divided into local fault diagnosis and centralized fault diagnosis. Local fault diagnosis takes place at power plant and substation facilities. Centralized fault diagnosis at the above level takes place at control centres equipped with EMS and SCADA systems using transmitted fault information.

In conventional systems, fault diagnosis is performed using a table of possible faults that contains information concerning operating protective relays, tripped circuit breakers, fault location, and fault type. When a fault occurs in the power system, this table is referred to in order to identify the location and type of fault.

This approach correctly diagnoses the case of a simple fault, like for example a single fault with correct operation of protective relays and circuit breakers. However, in the case of a single fault complicated by unwanted operation of protective relays and circuit breakers or simultaneous multiple faults as often occur in the case of lightning strokes, the processing becomes excessively complex and the diagnosis is not always correct.

When unwanted operation of devices and multiple faults are reflected in the fault diagnosis, the correct diagnosis can be obtained by performing various intersections of protection zones and other conditions. If, in addition, the installation of protective relays and sensors is inadequate, or the effects of weather and damage by animals are to be considered, knowledge based on experience becomes essential. If special relays are installed, or the power system is operated in a special network configuration, special processing may also be required.

3.2 Intelligent Fault Diagnosis

Due to the nature of fault diagnosis, an AI based approach has a particular role to play. Indeed, research into knowledge based fault diagnosis systems was the earliest attempt in AI applications to power systems. There are a large number of papers published and several systems already exist in actual operation. To a large extent, two major approaches have been adopted. The first approach consists of organizing monitoring information from operating relays and tripped circuit breakers during a fault and establishing a relationship to fault conditions into a tree structure in tabular form. This is referred to as the monitoring information based approach. In the second approach, the structure and functions of the protective relaying system are modelled, the fault conditions are simulated and a diagnosis is made by comparing the simulation results with the actual monitoring information. This is referred to as model-based approach. An overview of the two approaches can be found in reference [1].

In this section, an expert system for fault section estimation based on the first approach developed by Fukui et al [2] will be introduced. This system is written in Prolog. The system makes inferences to estimate the faulted section using conceptual and simplified knowledge and information in almost the same way as expert dispatchers do. The only knowledge and input information necessary are conceptual knowledge about protective relaying schemes and integrated and simplified signals from operating relays and tripped circuit breakers. The system answers the following questions:

(i) Which is the most likely section where the fault causing the present situation has occurred?

(ii) How did the fault cause the present situation? What was the sequence of the relay operations and the tripping of circuit breakers?

In this expert system, knowledge of the power system is stored in the databases as facts, and the relaying schemes are represented as rules. After information about operating relays and tripped circuit breakers are inputted, the inference engine estimates the faulted section which justifies each operation of the relays. The pattern matching and backtracking mechanism of Prolog checks stored facts and rules and finds the faulted section which satisfies each rule. This is the basic procedure for fault estimation using this technique.

Figure 3.1 shows the concept of the system. The inference engine controls and proceeds in the inference using rules and facts in three kinds of databases. Database 1 stores the facts on the network configuration and on the connections between all relays and circuit breakers in the power system. Database 2 involves the description of relaying schemes. Database 3 consists of the rules and heuristics by the expert dispatchers. These rules and heuristics control the inference and realize an efficient solution process.

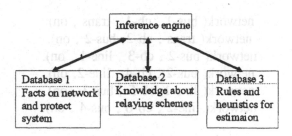

Figure 3.1 General structure of a fault section estimation expert system

3.2.1 Knowledge base

(1) Data base 1: Facts on network and protection system

This database stores knowledge on the configuration of the power system and on the connections between circuit breakers and protective relay systems, as well as the on-off-states of each circuit breaker.

The power system is represented by nodes and branches. In this paper, nodes represent sections which can be disconnected from other parts of the power system by opening circuit breakers, and branches represent circuit breakers. Figure 3.2 shows an example, where the original network configuration is shown in Figure 3.2(a), and its graphical representation, in Figure 3.2(b).

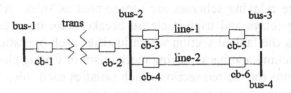

(a) Original network

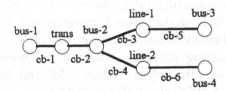

(b) representation by nodes and branches

network(bus-1 , cb-1 , trans , on).
network(trans , cb-2, bus-2 , on).
network(bus-2 , cb-3 , line-1 , on).
network(bus-2 , cb-4 , line-2, on).
network(line-1, cb-5 , bus-3 , on).
network(line-2 , cb-6 , bus-4 , on).

(c) Statements in Prolog

Figure 3.2 Representation of network configuration

This representation has the following advantages:

(i) A section represented by a node is a basic unit of the protective zone of relay systems. The protective zone of each relay corresponds to a set of nodes.

(ii) A black-out area is represented as a set of nodes which are disconnected from any node representing generator.

In the database, the graphical relation is described by a set of facts as described in the following statement.

network(<node name>, <CB name>,<node name>, on/off).   (3.1)

For example, the first line in Figure 3.2(c) indicates that the circuit breaker "cb-1" between the section "bus-1" and "trans" is closed at present. The network database is a set of those facts, and each fact is independent. When a circuit breaker is opened or closed, only a corresponding fact is changed.

(2) Database 2: Relaying schemes

(a) Reduction of relay signals

In an actual protection system, various types of relays are combined to make up a protection system. Usually, dispatchers recognize integrated signals from each protection system, not individual signals from component relays, i.e. dispatchers are not interested in relay signals at a detailed level, but on an integrated and simplified level. For example, in the case of a distance relay composed of mho relays and reactance relays, dispatchers usually recognize it as a single distance relay. Thus relays on this simplified and integrated level are called "reduced relays". Table 3.1 lists five common types.

(b) Statement of relays

The following statement is a general description of reduced relays.

relay($<$name$>$, $<$type$>$, $<$cbs$>$, $<$node$>$, $<$primary relay$>$, $<$backward nodes$>$).   (3.2)

where $<$name$>$ is the proper name of reduced relays., $<$type$>$ is its type name given in Table 3.1, $<$cbs$>$ is the set of circuit breakers which are tripped by this reduced relay, and $<$node. is the primary protective section if the relay type is one of MR, TR, or BR, $<$primary relay$>$ is necessary only when the relay type is LR, and is the proper name of the corresponding primary relay, $<$backward nodes$>$ is necessary only when the relay type is a type RR relay and it represents the backward nodes of protective direction.

Two examples are shown in Figure 3.3. The statement for rx-1 (Type TR: differential relay) and rx-2 (type RR: distance relay) in Figure 3.3 are

relay(rx-1, tr, [cb-1,cb-2], [trans-12], [], [])   (3.3)
relay(rx-2, rr, {cb-3], [], [], [bus-3])   (3.4)

In these statements, the elements surrounded by brackets "[" and "]" indicate a list, which is an ordered sequence of elements that can have any length. The notation "[]" is an empty list.

Table 3.1 Five types of reduced relays

Type	Explanation	Relaying Schemes
MR	Relays which use electric values obtained at two or more substations, and make the circuit breakers trip only in one substation. An example is differential reay for primary line protection.	Relay ... Relay / Line
TR	Relays for primary transformer protection.	Relay trans
BR	Relays for primary protection of buses. Usually differential relays are adopted for this kind of purpose.	Relay / Bus
RR	Distance relays, which use electric values in one station, and are often used as remote back-up relays.	Relay
LR	Local back-up relays which detect the operation failure of circuit breakers, and make the alternative circuit breakers trip.	Relay / Bus

Notes	☐ Closed circuit breakers	☒ FGailed-to-trip circuit breaker
	■ Tripped circuit breakers	

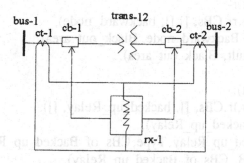

(a) Relaying scheme of type TR

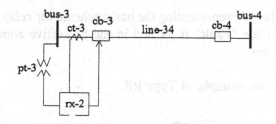

(b) Relaying scheme of type RR

Figure 3.3 Examples of relaying scheme

(c) Rules of relay operation

The rules representing relaying schemes are described as follows.

operate(Relay,Fault):-protect(Relay,Fault).   (3.5)

Protect(Relay,Fault):-
    relay(Relay,mr,CBs, zones, [],[]),
    member(Fault, Zones).   (3.6)

Protect(Relay,Fault):-
    relay(Relay,tr,CBs, zones, [],[]),
    member(Fault, Zones).   (3.7)

Protect(Relay,Fault):-
    relay(Relay,br,CBs, zones, [],[]),
    member(Fault, Zones).   (3.8)

Protect(Relay,Fault):-
      relay(Relay,rr,CBs, [],[], backward_node),
      search(CB, Backward_node, Black_out_area),
      member(Fault, Black_out_area).   (3.9)

Protect(Relay,Fault):-
      relay(Relay,lr,CBs, [], backed_up_Relay, []),
      operated(Backed_up_Relay),
      relay(backed_up_Relay,Type,CBs_of_Backed_up_Relay,Zones,{},{}),
      member(CB, CBs_of_Backed_up_Relay),
      not(tripped(CB)),
      member(Fault,Zones).   (3.10)

Statement (5) is the rule representing the basic scheme for relay operation in case
of a fault; is the node "Fault" is located in the protective zone of the "Relay",
the "Relay" operates.

Figure 3.4 shows an example of Type RR.

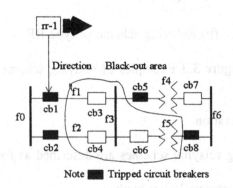

(a) Network with black-out area

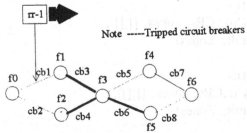

(b) Network representation with nodes and branches

Figure 3.4 Network representation with nodes and branches

3.2.2 Inference rules for fault section estimation

(1) States of relays and circuit breakers

Three states of relays and circuit breakers are discussed:

(i) "normal operation" and "normal tripping"
(ii) "false operation" and "false tripping"
(iii) "failure of operation" and "failure of tripping"

(2) The hierarchy of fault estimating rules

The rules for fault estimation are organised as a hierarchy as shown in Figure 3.5. The basic level consists of the inference rules corresponding to the single fault estimation. The upper levels consist of the rules for multiple fault estimation, the rules for estimating false operations of relays, and the special rules for some memorized fault patterns. The last rules are like know-how rules which expert dispatchers provide.

All these hierarchical rules are used dynamically. First, the fault is assumed to be a single fault. When any contradiction is proved or the problem has no answer under the hypothesis of single fault, the rules for multiple faults will be used. If there is any contradiction under the assumption of multiple faults, the existence of false operations are tested, and know-how rules for special fault patterns are used.

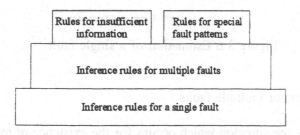

Figure 3.5 Hierarchy of inference rules

(3) Inference rules for single fault

Estimation of a single fault is performed by finding the node which justifies all the relay operations using Prolog's inference mechanism.

Consider a set of operating relays (ryi;i=1,2,...,n). If the fault is single and there is no false relay operation, only one node X satisfies all the predicate "operate(ryi,X)"(i=1,2,..,n). The predicate "operate" is defined as rule in the previous section.

By executing the following statement,

operate(ry1, X), operate(ry2, X), ..., operate(ryn, X).   (3.11)

The Prolog system checks each rule and substitutes the fault section into the variable "X".

Figure 3.6 conceptualizes this process. For each operating relay, one or more candidates for fault sections are obtained by using each rule. each circle Fi indicates a set of these candidates, and their intersection indicates the real fault section X.

The inference rule for single fault does not always give correct answers, because there may be failures in relay operation or false operation of relays.

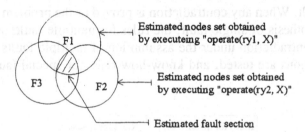

Figure 3.6 Estimation of a single fault

(4) Inference rules for multiple faults

Figure 3.7 shows the situation which occurs for the existence of more than one fault. When all the relays operate normally, local intersections appear which correspond to each fault.

The inference rules for multiple faults are used to divide the operating relays into groups corresponding to each single fault. Then the inference rule for a single fault is applicable within these divided groups and gives fault sections which correspond to each local intersection.

The division of the operating relays is also performed for the existence of falsely operated relays. Figure 3.8 shows a case in which a relay operates falsely. In this case, the node set determined by the falsely operating relay is isolated from the others, but there is no difference between the node set determined by normally operating relays and those by falsely operated relays. Therefore it is rather difficult to distinguish falsely operating relays from normally operating ones. As mentioned, the inference engine regards all the operations of relays as normal ones in the first step. The inference engine finds the false operation by recognizing the special patterns in the information of the operating relays and on-off-states of the circuit breakers. If there are no special patterns, it concludes that there is no false operation.

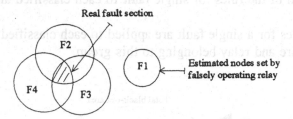

Figure 3.7. Estimation of multiple faults

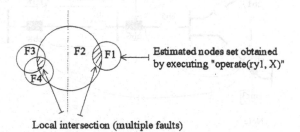

Figure 3.8 Estimation with a falsely operating relay

(i) Division of black-out area

The fault sections must be within the black-out area. In the case of multiple faults, these areas may be divided into sub-areas which are electrically independent from each other. For example, in Figure 3.9, "cb6" divides the total black-out area into sub-areas 1 and 2. The sub-areas are calculated by almost the same process that checks for existence of the electric paths between two nodes, and the process is executed by the predicate "search" described in statement (3.9).

(ii) Classification of operating relays

Next, the operating relays are classified into groups corresponding to sub-areas of a black-out area. If a sub-area includes a node which is a member of the protective zone of a relay, the relay is regarded as belonging to the group corresponding to this sub-area. In Figure 3.9, the protective zones of "rr1", "rr6" (type RR) and "mr3" (type MR) belong to sub-area 1, and those of "mr8" belong to sub-area 2. So, "rr1", "rr6" and "mr3" are classified into group 1, corresponding to sub-area 1, and "mr8" is classified into group 2, corresponding to sub-area 2.

(iii) Application of the rules for single fault to each classified area

Finally, the rules for a single fault are applied to each classified group for each black-out sub-are and relay belonging to this group.

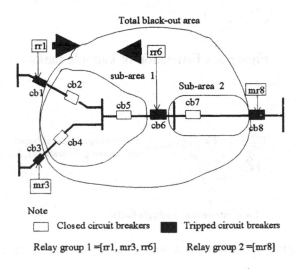

Figure 3.9 Separation of a black-out

(5) Rules for finding failure of operation

The existence of relay operation failures causes a lack of information, and sometimes the estimation gives several candidates for fault sections. In this case, the rules for finding the failure of relay operations and the tripping failure of circuit breakers are invoked. Some examples of these rules are given in the following.

(i) Rules

(rule 1) If a relay operated and the circuit breaker which receives the tripping signal from the relay has not tripped, then the circuit breaker can be considered as fail to trip.

(rule 2) If an estimated fault section belongs to protective zone of an unoperating primary relay, then the primary relay can be considered as fail to trip.

(rule 3) If a type RR relay which did not operate belongs to a group corresponding to a black-out sub-area, and if the protective zone of the relay includes the estimated fault section, then the relay can be considered as fail to operate.

(ii) Counting the number of failed relays and circuit breakers

Figure 3.10 gives an example. Only two RR relays on both ends operated, and there are three possible fault sections, [f1,f2,f3].

By using the above mentioned rules, the inference engine counts the number of the failed relays and circuit breakers. The following is a list of failures of operations in Figure 10.

If "f1" is the fault section;
    "mr1" and "mr2" (primary line protection) failed.
If "f2" is the fault section;
    "br2" (primary bus protection) failed.
If "f3" is the fault section;
    "mr3" and "mr4" (primary line protection) failed.
    "rr3" (remote back-up) failed.

By counting the number of failures of relay operation and circuit breaker tripping, the inference engine determines the most possible fault section by considering the number of failures.

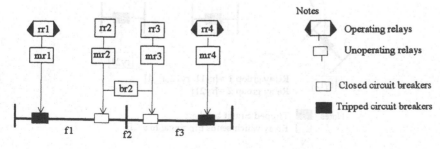

Figure 3.10 Three possibilities of fault section

(6) Rules for finding possible situations with false operation

In a multiple fault case, especially when a black-out area is divided into sub-areas, we can sometimes recognize special patterns in the on-off-states of circuit breakers and relay operations which belong to each sub-area. This special pattern is called the "inter-relation between two sub-areas" and additional inference rules are introduced to deal with it.

Figure 3.11 shows an example for the interrelation between two black-out sub-areas. There is one tripped circuit breaker "cb1" between them. "cb1" is tapped by the relay "rx-1" which belongs to group 1 corresponding to sub-area 1.

Under the conditions mentioned above, there are several possible explanations.

(Explanation 1)
Two faults occurred individually in sub-area 1 and 2. Although the relays belonging to each group operated normally, the relay which belongs to group 2 and should make "cb1" trip, failed to operate or was blocked to operate.

(Explanation 2)
The fault occurred within sub-area 1, and relays of group 1 operated normally, while all the relays of group 2 operated falsely.

(7) Sequence of relay and circuit breaker operation

After estimating the fault section, the sequence of relay operations and tripping can be estimated using the basic rules of relaying schemes. Using this function, the inference system can explain how the faults created the present situation.

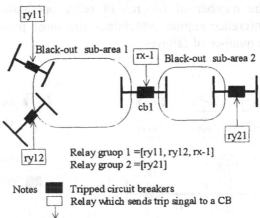

Relay group 1 =[ry11, ry12, rx-1]
Relay group 2 =[ry21]

Notes   ▮ Tripped circuit breakers
        ☐ Relay which sends trip singal to a CB

Figure 3.11 Interrelation between two sub-areas of a black-out area

3.3 Example Implementations

The system described in section 3.2 was implemented by Hitachi of Japan. In the UK, the real time Alarm Handing and Fault Analysis (AHFA) expert system project began under the CEGB in 1989. The AHFA expert system was first developed in 1990, demonstrated as a small scale prototype in 1991, then scaled up in 1992 to a full area network and tested for extended periods at an Area Control Centre. AHFA accepts switching indication messages from a file or serial data port and produces on-line diagnosis summaries of transient or permanent faults on components from the corresponding set of messages. The on-line summaries are displayed in an interactive window where they can be expanded into a number of more detailed page-size reports, for example including messages indirectly relevant to and preceding the diagnosis.

References

1. Y Sekine, Y Akimoto, M Kunugi, C Fukiu, S Fukui, Fault diagnosis of power systems, Procs of the IEEE, Vol.80, No.5, 1992
2. C Fukui, J Kawakami, An expert system for fault section estimation using information from protective relays and circuit breakers, IEEE Trans Power Delivery, Vol.PWRD-1, No.4, 1986, pp.83 - 90
3. E Cardozo, S N Talukdar, A distributed expert system for fault diagnosis, IEEE Trans Power Systems, Vol.3, No.2, 1988, pp.641 - 646
4. IEEE Working Group Report, Potential applications of expert systems to power system protection, IEEE Trans Power Delivery, Vol.9, No.2, 1994, pp.720 - 728
5. C C Liu, A V Son, Expert systems: development experience and user requirements, Electra No. 146, 1993, pp.29 - 67
6. D G Esp, A O Ekwue, J F Macqueen, B M Vaughan, AHFA - a real-time expert system for the incremental diagnosis of multiple faults on a transmission network using the sequence and timing of switching indications, Proc IEE Control'94, pp.141 - 145

CHAPTER 4
DECISION SUPPORT FOR REACTIVE POWER AND VOLTAGE CONTROL

4.1 Reactive Power and Voltage Control in Power Systems

Reactive power and voltage control plays an important role in the security of the modern power systems. For efficient and reliable operation, the control of reactive power and voltage should satisfy the following objectives:

(1) The voltages at the terminals of all equipment in the system should be within acceptable limits. Both utility equipment and customer equipment must be designed to operate at a certain voltage rating because prolonged operation of the equipment at voltages outside the allowable ranges could adversely affect performance and cause damage to it.

(2) System stability is enhanced to maximize utilization of the transmission system. As observed by utilities worldwide, voltage collapse is a hidden threat to the secure operation of modern power systems.

(3) The reactive power flow is minimized so as to reduce active and reactive losses to a practical minimum. This ensures that the transmission system operates efficiently, i.e., mainly for active power transfer.

The problem of maintaining voltages within the required limits is complicated by the fact that the power system supplies power to a vast number of loads and is fed from many generating units. As loads vary, the reactive power requirements of the transmission system vary. Since reactive power cannot be transmitted over long distances, voltage control has to be effected by using special devices dispersed throughout the system. This is in contrast to the control of frequency which depends on the overall system active power balance. the proper selection and coordination of equipment for controlling reactive power and voltage are among the major challenges of power system engineering.

The control of voltage levels is accomplished by controlling the production, absorption, and flow of reactive power at all levels in the system. The generating units provide the basic means of voltage control; the automatic voltage regulators control field excitation to maintain a scheduled voltage level at the terminals of the generators. Additional means are usually required to control voltage throughout the system. The devices used for this purpose may be classified as follows:

(1) Sources or sinks of reactive power, such as shunt capacitors, shunt reactors, synchronous condensers, and static var compensators.
(2) Line reactance compensators, such as series capacitors.
(3) Regulating transformers, such as tap-changing transformers and boosters.

4.2 An Expert System for Reactive Power and Voltage Control

4.2.1 The expert system configuration

Figure 4.1 illustrates an expert system configuration for reactive power and voltage control, which is similar to the general structure of the expert system shown in Figure 1.1. The proposed rules are separated into two depending on the functions they perform. Rule base 1 performs the detection of voltage problems and the search for control actions based on empirical knowledge. When it is judged that the problem is serious enough such that empirical judgment may not be reliable, Rule Base 2 aids the user in formulating the problem in order to utilize an available software package which provides a more systematic analysis.

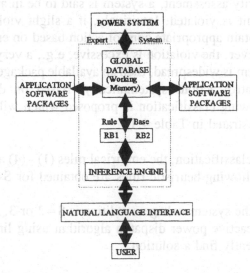

Figure 4.1 General structure of an expert system for voltage control

4.2.2 The knowledge base

A successful expert system relies on a high-quality knowledge base. The knowledge base needs to incorporate the knowledge (facts, heuristics) required to perform a task well. For the reactive power/voltage control problem, the following empirical rules are identified:

(1) If a load bus voltage drops below or rises above the operating limit, control devices such as shunt capacitors, transformer tap changers, generator voltages and synchronous condensers can be switched or adjusted to restore the voltage profile.

(2) If a load bus voltage drops below rises above the operating limit, it is most efficient to apply the reactive compensation locally. If the capacity of local compensators is insufficient to resolve the voltage problem, then compensators with the next highest sensitivities should be chosen.

(3) If the voltage is low or high at a bus, the tap position of the local transformer tap changer can be raised or lowered to correct the problem. However, increasing or decreasing the tap position may cause other load bus voltages to drop or rise.

(4) Generator bus voltages can be raised or lowered to solve the low or high load voltage problem.

In steady-state security assessment, a system is said to be in an emergency state if any operating limit is violated. Intuitively, if a slight violation occurs, one should be able to obtain appropriate control action based on empirical rules (2), (3) and (4). If, however, the violation is excessive, e.g., a very low bus voltage, or the voltage problem is widespread, then an available package will be necessary to generate the control actions systematically. Based on different levels of violations, the following classification is proposed which will be denoted by a severity index S illustrated in Table 4.1.

Based on the above classification, the empirical rules (1) - (4) apply for problems where $S = 1$. The following heuristic rules are obtained for $S = 2$ or 3.

(5) If the index for the system voltage condition is $S = 2$ or 3, a more systematic method such as a reactive power dispatch algorithm using linear programming is required to efficiently find a solution.

(6) If the index for the system voltage condition is $S = 2$, the objective of the

reactive power dispatch algorithm should be to minimize the number of control devices involved.

Table 4.1 Definition for S index

S	Operating Condition	Description of the Problem
0	Normal	All voltage limits satisfied
1	Emergency	Slight violation of the voltage limit at one or two buses
2	Emergency	Severe violation of one voltage limit or three or more bus voltage limits violated, and local and neighbouring control devices are sufficient to solve the problem
3	Emergency	Same as above, except local and neighbouring control devices are not sufficient to solve the problem

(7) If the index for the system voltage condition is $S=3$, the reactive power dispatch algorithm should be utilized to find a feasible solution as quickly as possible. Other objectives have lower priorities at this time. If necessary, emergency bus voltage limits should be used.

4.2.3 The data structure

The general form of the data representation is given by type, name, status, attributes and characteristics.

(1)   Load buses
      (name, location, status, MW demand, MVAR demand, voltage V, V-upper limit, V-lower limit)

(2)   Reactive controllers -- tap settings, reactive injections, generator voltages
      (name, type, location, status, current value, upper limit, lower limit)

4.2.4 Production rules

For reactive power management, 28 rules based on the knowledge are described in Section 4.2.2. Rule Base 1 concerns itself with less severe voltage problems and attempts to correct the voltage problem with a small amount of computation by utilizing the first twelve rules listed below. Sensitivities for reactive compensators, generator voltage control, and transformer tap changers derived in Section 4.2.3, are incorporated into the Rule Base 1 rules to obtain the best estimates of the control required to restore the voltage profile. Rule Base 2 formulates decisions for more severe voltage problems by utilizing a linear programming package. In these rules, sensitivities and available control capacities are used to classify the severity of the voltage problem. Voltage problems of severity rating 3 are solved using emergency constraints with the cost function for the Linear Programming package set to zero. Severity rating 2 problems are solved with normal constraints and the objective is to minimize the weighted sum of the control adjustments.

(1) Rules for Rule Base 1

Rule 1 -- If a load bus voltage is below voltage limits and above emergency voltage limits then identify bus as having a low voltage.

Rule 2 -- If a load bus voltage is above voltage limits and below emergency voltage limits then identify bus as having a high voltage.

Rule 3 -- if a problem bus voltage is now within limits then clear voltage control and verify with load flow.

Rule 4 -- If the voltage violation is a low voltage and the compensator to be used for control is available then compute maximum voltage increase.

Rule 5 -- If the voltage violation is a high voltage and the compensator to be used for control is available then compute maximum voltage decrease.

Rule 6 -- If the compensator can provide the necessary control than implement the necessary compensation.

Rule 7 -- If control compensator is not at limits and compensator is insufficient to eliminate voltage violation then move compensator to limit.

Rule 8 -- If the compensator to be used for control has not been chosen then identify voltage controller.

Rule 9 -- If the control compensator is at full output and there remain controllers to be checked then select next controller.

Rule 10 -- If the control compensator is not available and there remain controllers to be checked then select next controller.

Rule 11 -- If the control compensator is not available and there are no more controllers to be checked then severity rating = 3.

Rule 12 -- If the control compensator is at full output and there are no more controllers to be checked then severity rating = 3.

(2) Rules for Rule Base 2

Rule a1 -- If more than two load bus voltages are below voltage limits then initiate determination of problem severity.

Rule a2 -- If more than two load bus voltages are above voltage limits then initiate determination of problem severity.

Rule a3 - If both high and low voltage violations are present then severity rating is 3.

Rule a4 -- If a load bus voltage is outside voltage emergency limits then initiate determination of problem severity.

Rule a5 -- If the n most significant (sensitive) controllers have been determined to be insufficient to alleviate the voltage problems then severity rating is 3.

Rule a6 -- If the n most significant controllers have been determined to be sufficient to alleviate the voltage problems then severity rating is 2.

Rule a7 -- If a load bus voltage is below voltage limits and has not been checked for available control then determine magnitude of voltage violation.

Rule a8 -- If a load bus voltage is above voltage limits and has not been checked for available control then determine magnitude of voltage violation.

Rule a9 -- If a low voltage exists and the controller under consideration is available then calculate voltage control available from this controller.

Rule a10 -- if a high voltage exists and the controller under consideration is available then calculate voltage control available from this controller.

Rule a11 -- If the first controller on the list of controllers to be checked is unavailable then select next controller.

Rule a12 -- If severity rating=2 then run LP problem using normal constraints and unity weighting factor for all control adjustments.

Rule a13 -- If severity rating=3 then run LP problem using emergency constraints and feasibility only (set cost function to zero).

Rule a14 -- If LP cannot resolve the voltage violations using normal constraints than increase the severity rating to 3.

Rule a15 -- If LP cannot resolve the voltage violations using emergency constraints then inform operator that system cannot be kept within emergency limits using nondisruptive controls.

Rule a16 -- If all voltages are within voltage emergency limits then clear voltage emergency.

For the inference engine, the forward reasoning was adopted.

4.3 Example Implementations

Based on the proposed technique of section 4.2, a prototype system was implemented for Electicdade de Portugal. A similar on-line application has been on-going for the National Grid Company (NGC) of the UK to develop a knowledge based system for voltage monitoring and control. In this application the Optimal Power Flow calculates an hourly voltage schedule and the rule-based system suggests remedial actions whenever voltage violation alarms are detected by the state monitoring which follows every state estimation.

References

1. C C Liu, K Tomsovic, An expert system assisting decision-making of reactive power/voltage control, IEEE Trans Power Systems, Vol.PWRS-1, No.3, 1986, pp.195 - 201

2. L M F Barruncho, J P Sucena Paiva, C C Liu, R Pestana, A Vidigal, Reactive management and voltage monitoring and control, Electrical Power and Energy systems, Vol.14, 1992, pp.144 - 157

3. S J Cheng, O P Malik, G S Hope, An expert system for voltage and reactive power control of a power system, IEEE Trans Power Systems, Vol.3, No.4, 1988, pp.1449 - 1455

4. L M Barruncho, C C Liu, A O Ekwue, D G Esp, J F Macqueen, B W Vaughan, Application of a knowledge based system for voltage monitoring and control on the NGC system, Proc IEE Control'95, pp.146 - 152

5. Y H Song, J Macqueen, D T Y Cheng, A Arnal, On voltage stability in electric power systems, Proc IEEE/CSEE International Conference on Power System Technology, 1994, pp.1236 - 1239

CHAPTER 5
FUZZY LOGIC: AN INTRODUCTION

5.1 Characteristics of Fuzzy Logic Systems

Experts often rely on common sense to solve problems. The knowledge an expert uses is often vague and ambiguous. Fuzzy logic attempts to model computer reasoning on the kind of imprecision and undecidability found in human reasoning. Through fuzzy logic, a system cannot only represent such imprecise concepts as fast, stable, and inexpensive, but through a set of sound mathematical principles, it can also use these concepts to make deductions about a system.

5.1.1 Fuzzy set

(1) Linguistic variables

Fuzzy logic is primarily concerned with quantifying and reasoning about vague or fuzzy terms that appear in our natural language. In fuzzy logic, these fuzzy terms are referred to as linguistic variables (also called fuzzy variables). For example, in the statement " The speed of the generator is fast", the speed is a linguistic variable with its value being fast. We call the range of possible values of a linguistic variable, the variable's universe of discourse.

(2) Fuzzy set representation

First recall traditional set theory which views the world as black or white. That is, a crisp set A in a universe of discourse U can be defined by listing all of its members or by identifying the elements $x \subset A$. One way to do the latter is to specify a condition by which $x \subset A$; thus A can be defined as $A = \{x \mid x$ meets some condition$\}$. Alternatively, we can introduce a zero-one membership function for A, denoted $\mu_A(x)$, such that $A \Rightarrow \mu_A(x) = 1$ if $x \subset A$ and $\mu_A(x) = 0$ if $x \not\subset A$.

A fuzzy set F defined on a universe of discourse U is characterized by a membership function $\mu_F(x)$ which takes on values in the interval [0, 1]. A fuzzy

set is a generalization of a crisp subset whose membership function only takes
on two values, zero or unity. A membership function provides a measure of the
degree of similarity of an element in U to the fuzzy subset.

Figure 5.1 illustrates fuzzy and crisp sets of "young" people.

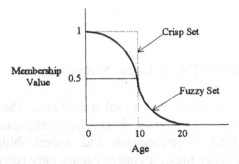

Figure 5.1 The fuzzy and crisp sets of "young" people

A fuzzy set F in U may be presented as a set of ordered pairs of generic element
x and its grade of membership function: $F=\{(x, \mu_F(x))|x \in U\}$. When U is
continuous, F is commonly written as $F= \int_U \mu_F(x)/x$. In this equation, the
integral sign does not denote integration; it denotes the collection of all points
$x \in U$ with associated membership function. When U is discrete, F is commonly
written as $F=\Sigma_U \mu_F(x)/x$. The slash in these expressions associates the elements
in U with their membership grades.

(3) Membership functions

In engineering applications of fuzzy logic, the most commonly used shapes for
membership functions are triangular, trapezoidal, piecewise linear and Gaussian.
Until very recently, membership functions were arbitrarily chosen by the user,
based on the user's experience; hence, the membership functions for two users
could be quite different. The number of membership functions is dependent on
the users and greater resolution is achieved by using more membership functions
at the price of greater computational complexity.

(4) Hedges

A linguistic hedge or modifier is an operation that modifies the meaning of a
term, or more generally, of a fuzzy set. For example, if weak pressure is a fuzzy
set, then very weak pressure, more-or less weak pressure, extremely weak
pressure, and not-so weak pressure are examples of hedges which are applied to

this fuzzy set. Hedges can be viewed as operators that can act upon a fuzzy set's membership function to modify it. Five commonly used hedges shown in Figure 5.2 are introduced in the following:

Concentration (very)

The concentration operation has the effect of further reducing the membership values of those elements that have smaller membership values. This operation is given as:

$$\mu_{CON(U)}(x) = (\mu_U(x))^2 \tag{5.1}$$

If, for example, weak pressure has a membership function μ, the very weak pressure is a fuzzy set with membership function μ^2, and very very weak pressure is fuzzy set with membership function μ^4.

Dilation (somewhat)

The dilation operation dilates the fuzzy elements by increasing the membership value of those elements with small membership values more than those elements with high membership values. This operation is given as:

$$\mu_{dil(U)}(x) = (\mu_U(x))^{1/2} \tag{5.2}$$

Intensification (indeed)

the intensification operation has the effect of intensifying the measuring of the phrase by increasing the membership values above 0.5 and decreasing those below 0.5. This operation is given as:

$$\mu_{INT(U)}(x) = 2(\mu_U(x))^2 \quad \text{for } 0 \le \mu_{(U)}(x) \le 0.5$$
$$= 1 - 2(1 - \mu_U(x))^2 \quad \text{for } 0.5 < \mu_{(U)}(x) \le 1 \tag{5.3}$$

Artificial hedge

Two hedges that are quite useful are the plus and minus hedges. These artificial hedges provide milder degrees of concentration and dilation than those associated with the concentration and dilation hedges. The operations are given as:

$$\mu_{plus(U)}(x) = (\mu_U(x))^{1.25} \tag{5.4}$$
$$\mu_{minus(U)}(x) = (\mu_U(x))^{0.75} \tag{5.5}$$

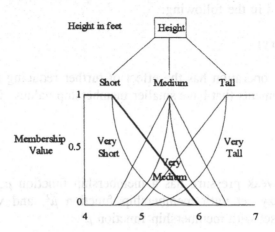

Figure 5.2 Membership functions

(5) Set theoretic operations

In fuzzy logic, union, intersection and complement are defined in terms of their membership functions. Let fuzzy sets A and B be described by their membership functions.

fuzzy union

$$\mu_{A \vee B}(x) = \max(\mu_A(x), \ \mu_B(x))$$
$$= \mu_A(x) \vee \ \mu_B(x) \tag{5.6}$$

fuzzy intersection

$$\mu_{A \wedge B}(x) = \min(\mu_A(x), \ \mu_B(x))$$
$$= \mu_A(x) \wedge \ \mu_B(x) \tag{5.7}$$

fuzzy complement

$$\mu_{\sim B}(x) = 1 - \mu_B(x) \tag{5.8}$$

5.1.2 Fuzzy logic vs. traditional logic systems

Fuzzy logic may be viewed as an extension of multivalued logic. Its uses and objectives, however, are quite different. Thus, the fact that fuzzy logic deals with approximate rather than precise modes of reasoning implies that, in general, the chains of reasoning in fuzzy logic are short in length, and rigor does not play as important a role as it does in classical logical systems. In a nutshell, in fuzzy logic everything, including truth, is a matter of degree. The main features of fuzzy logic that differentiate it from traditional logic systems are the following:

(1) In two-valued systems, a proposition p is either true or false. In multivalued logical systems, a proposition may be true or false or have an intermediate truth value, which may be an element of a finite or infinite truth value set T. In fuzzy logic, the truth values are allowed to range over the fuzzy subsets of T. For example, if T is the unit interval, than a truth value in a fuzzy logic, for example, "very true," may be interpreted as a fuzzy subset of the unit interval. In this sense, a fuzzy truth value may be viewed as an imprecise characterization of a numerical truth value.

(2) The predicates in two-valued logic are constrained to be crisp in the sense that the denotation of a predicate must be a nonfuzzy subset of the universe of discourse. In fuzzy logic, the predicates may be non-crisp - for example, "mortal," "even," and "father of" - or, more generally, fuzzy -for example, "tired", "large", "much faster" and "friend of."

(3) Two-valued as well as multivalued logics allow only two quantifiers: "all" and "some." By contrast, fuzzy logic allows, in addition, the use of fuzzy quantification exemplified by "most," "several," "few," "occasionally," and so on. Such quantification may be interpreted as fuzzy numbers that provide an imprecise characterization of the cardinality of one or more fuzzy or nonfuzzy sets. In this perspective, a fuzzy quantifier may be viewed as a second-order fuzzy predicate.

(4) Fuzzy logic provides a method for representing the meaning of both nonfuzzy and fuzzy predicate-modifiers exemplified by "not," "very," "more or less," "extremely," "slightly," and so on. This, in turn, leads to a system for computing with linguistic variables.

(5) In two-valued logical systems, a proposition p may be qualified, principally by associating with p a truth value, "true" or "false"; a modal operator such as "possible" or "necessary"; and an intensional operator such as "know" or "believe." Fuzzy logic has three principal modes of qualification:

truth-qualification, as in (Mary is young) is not quite true.
probability-qualification, as in (Mary is young) is unlikely.

possibility-qualification, as in (Mary is young) is almost impossible.

5.2 Fuzzy Logic Systems

Figure 5.3 depicts a fuzzy logic system that is widely used in fuzzy logic controllers and signal processing applications. A fuzzy logic system maps crisp inputs into crisp outputs. It contains four components: fuzzifier, rules, inference engine, and defuzzifier.

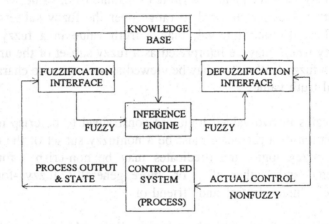

Figure 5.3 The general structure of a fuzzy logic system

5.2.1 Fuzzification

The fuzzifier maps a crisp point $x = col(x1,...,xn) \in U$ into a fuzzy set A in U. The most widely used fuzzifier is the singleton fuzzifier which is nothing more than a fuzzy singleton, i.e, A is a fuzzy singleton with support x' if $\mu_A(x') = 1$ for $x = x'$ and $\mu_A(x') = 0$ for all other $x \in U$ with $x \neq x'$.

Singleton fuzzification may not always be adequate, especially when data is corrupted by measurement noise. Nonsingleton fuzzification provides a means for handling such uncertainties totally within the framework of fuzzy logic system's. A nonsingleton fuzzier is one for which $\mu_A(x') = 1$ and $\mu_A(x)$ decreases from unity as x moves away from x'. In nonsingleton fuzzification, x' is mapped into a fuzzy number, i.e., a fuzzy membership function is associated with it. Examples of such membership functions are the Gasussian and triangular. The broader these functions are, the greater is the uncertainty about x'.

5.2.2 Fuzzy rules

A fuzzy system is characterized by a set of linguistic statements based on expert knowledge. The expert knowledge is usually in the form of "if-then" rules, which are easily implemented by fuzzy conditional statements. Furthermore, several linguistic variables might be involved in the antecedents and the conclusions of these rules. When this is the case, the system will be referred to as a multi-input-multi-output fuzzy system.

Fuzzy logic treats a fuzzy set as a fuzzy proposition. A fuzzy proposition is a statement that asserts a value for some given linguistic variable. The general form is

Proposition: X is A

where A is a fuzzy set on the universe of discourse X. A fuzzy rule relates two fuzzy propositions in the form:

IF X is A THEN Y is B

This rule establishes a relationship or association between the two propositions. Fuzzy logic systems store rules as fuzzy associations. That is, for the rule IF A THEN B, where A and B are fuzzy sets, a fuzzy logic systems stores the association (A,B) in a matrix M. The fuzzy associative matrix M maps fuzzy set A to fuzzy set B. This fuzzy association or fuzzy rules is called a fuzzy associative memory (FAM).

5.2.3 Fuzzy inference

In the fuzzy inference engine fuzzy logic principles are used to combine fuzzy IF-THEN rules from the fuzzy rule base into a mapping from fuzzy input sets to fuzzy output sets.

(i) Max-min inference

In max-min inference the implication operator used is min. That is:

$$m_{ij} = truth(a_i \rightarrow b_j) = min(a_i, b_j)$$

Given two fuzzy sets A and B, the above equation can be used to form the matrix M. It is illustrated in Figure 5.4.

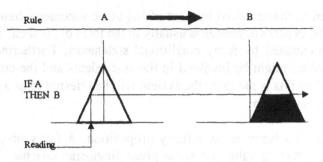

Figure 5.4 Max-min inference

(ii) Max-product inference

Max-product inference uses the standard product as the implication operator when forming the components of M:

$$m_{ij} = a_i b_j \qquad\qquad\qquad (5.9)$$

Following the calculation of this matrix, max-min composition is used to determine the induced matrix B' from some subset vector A'. It is illustrated in Figure 5.5.

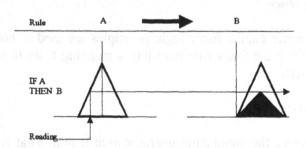

Figure 5.5 Max-product inference

5.2.4 Defuzzification

A defuzzifier produces a crisp output for our fuzzy logic system from the fuzzy set that is the output of the inference block. Many defuzzifers have been proposed in the literature; but, there are no scientific bases for any of them; consequently, defuzzification is an art rather a science. Because we are interested in engineering applications of fuzzy logic, one criterion for the choice of a defuzzifier is computational simplicity. This criterion has led to the following candidates for defuzzifiers.

(1) Maximum defuzzifier
This defuzzifier examines the fuzzy set B and chooses as its output the value of y for which is a maximum.

(2) Mean of maxima defuzzifier
This defuzzifier examines the fuzzy set B and first determines the values of y for which is a maximum. It then computes the mean of these values as its output.

(3) Centroid defuzzifier
This defuzzifier determines the center of gravity of B and uses this value as the output of the FLS. In the case of a discrete universe, this method yields:

$$y = [\sum_{i=1}^{I} \cdot y_i \mu_B(y_i)] / [\sum_{i=1}^{I} \mu_B(y_i)] \tag{5.10}$$

5.3 Fuzzy Logic in Power Systems

There are several major tasks for building a fuzzy logic expert systems. These are: (1) define the problem; (2) define the linguistic variables; (3) define the fuzzy sets; (4) define the fuzzy rules; (5) build the system; (6) test the system and (7) tune the system. Fuzzy set theory has been applied to a wide range of power system problems, including: (1) power system stability control; (ii) power system stability assessment; (iii) power system optimization; (iv) power system protection etc.

References

1. J M Mendel, Fuzzy logic systems for engineering: a tutorial, Proc of the IEEE, Vol.83, No.3, 1995, pp.345 - 377
2. C C Lee, Fuzzy logic in control systems: fuzzy logic controller - Part I & Part 2, IEEE Trans Systems, Man and Cybernetics, Vol.20, No.2, 1990, pp.404 - 435
3. J Durkin, Expert systems: design and development, Macmillan Publishing Company, 1994

CHAPTER 6
FUZZY STABILITY CONTROL FOR POWER SYSTEMS

6.1 Power System Stability Control

The continuing interconnection of bulk power systems and increasing utilization of existing facilities, brought about by economic and environmental pressures, has led to an increasingly complex power system that is being pressed to operate ever closer to the limits of the system. The operating environment has contributed to the growing importance of problems associated with all aspects of the stability of power systems. A major concern for power engineers is to maintain adequate control at all times. In recent years the utilisation of additional/supplementary control measures for improving transient and dynamic stability has received much attention. In particular, power system stabiliser (PSS) design techniques continue to take advantage of the state-of-the-art control theories and the developments in hardware technology. PSSs based on the eigenvalue analysis methods, where the stabiliser loop parameters are chosen to counteract some critical system modes, have been applied to many commercial power systems. The application of linear optimal control theory for the purpose of deriving a constant feedback gain matrix for the stabiliser loop has been extensively studied. During the past decade, considerable efforts have been made to develop adaptive and intelligent control for power systems. Fuzzy logic, as a promising branch of artificial intelligence, has been successfully employed in engineering fields. It is a method of easily representing knowledge/number of uncertainty and ambiguity on a digital computer. In the following section, a fuzzy logic power system stabiliser proposed in reference [1] will be described to demonstrate how fuzzy logic can be applied to power system control design.

6.2 Fuzzy Logic Stability Control

The basic structure of a fuzzy logic power system stabiliser is similar to Figure 5.3. The principal design parameters for a fuzzy PSS are: (i) fuzzification strategies; (ii) rule base; (iii) inference engine and (iv) defuzzification strategies. These will be discussed in this section.

6.2.1 Fuzzifier - specify the membership functions for stabiliser inputs

To make the PSS capable of providing the desired system damping under disturbance conditions, state variables representative of system dynamic performance must be taken as the input signal to the fuzzy PSS. Selection of the control variables relies on the nature of the system and its desired output. As the target of the application is to stabilise and improve the damping of a synchronous machine, speed deviation $\Delta\omega$ and active power deviation ΔPe are selected as controller inputs.

Each of the FLC input and output signals is interpreted into a number of linguistic variables. The number of linguistic variables varies according to the application. Usually an odd number is used. Each linguistic variable has its fuzzy membership function. The membership function maps the crisp values into fuzzy variables. A set of membership defined for seven linguistic variables negative big (NB), negative medium (NM), negative small (NS), zero (Z), positive small (PS), positive medium (PM) and positive big (PB), respectively, is shown in Figure 6.1.

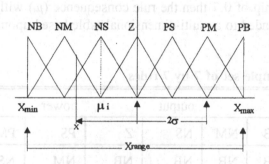

Figure 6.1 Membership functions for seven linguistic variables

For simplicity, it is assumed that the membership functions are symmetrical and each one overlaps the adjacent functions by 50%. Thus the controller input and output parameters (herein called gains), can be defined as

$$K_j = 2/(X_{maxj} - X_{minj}) \qquad j = 1, 2, 3 \tag{6.1}$$

where X_{maxj} and X_{minj} are the maximum and the minimum values of the control variables X_j. The input and output gains, K_j, are referred to as the FLC parameters.

6.2.2 Rule base and inference engine

In general, a fuzzy system maps input to output fuzzy sets. Fuzzy rules are the relations between input/output fuzzy sets. They are usually in the form if A, then B, where A is the rule and defines a fuzzy patch in the cartesian product $A \times B$. The antecedents of each fuzzy rule describe a fuzzy input region in the state space.

For a system of two control variables with seven linguistic variables in each range, this leads to a 7×7 decision table as shown in Table 6.1. Every entity in the table represents a rule. The antecedent of each rule conjoins the speed deviation or error and the deviation in the generated power fuzzy set values. An example of, say, the ith rule is

If $\Delta \omega$ is NB and ΔP is NM then U is NB

Using min-max inference, the activation of the ith rule consequently is a scalar value (μ_i) which equals the minimum of the two antecedent conjunct' values. For example if $\Delta \omega$ belongs to NB with a membership of 0.3 and ΔPe belongs to NM with a membership of 0.7 then the rule consequence (μ_i) will be 0.3. The rules table can be extended to a multi-dimensional table based upon the number of the control variables.

Table 6.1 A sample set of 7 by 7 rules

speed	output				power		deviation
deviation	NB	NM	NS	Z	PS	PM	PB
NB	NB	NB	NB	NB	NM	NS	Z
NM	NB	NB	NM	NM	NS	Z	PS
NS	NB	NM	NM	NS	Z	PS	PM
Z	NM	NM	NS	Z	PS	PM	PM
PS	NM	NS	Z	PS	PM	PM	PB
PM	NS	Z	PS	PM	PM	PB	PB
PB	Z	PS	PM	PB	PB	PB	PB

The knowledge required to generate the fuzzy rules can be derived from an offline simulation, an expert operator and/or a design engineer. Some knowledge can be based on the understanding of the behaviour of the dynamic system under control. Normally, rule definition is based on the operator's experience and the engineer's knowledge. However, it has been noticed in practice that, for monotonic systems, a symmetrical rule table is very appropriate, although sometimes it needs slight adjustment based on the behaviour of the specific system. If the system dynamics are not known or are highly nonlinear, trial-and-error procedures and experience play an important role in defining the rules.

6.2.3. Defuzzifier - output

Once the membership values for the stabiliser output have been computed, a suitable algorithm must be employed to determine the stabiliser output signal. There are several such algorithms introduced in Chapter 5. If the centroid defuzzifier is employed, a discretised output universe of discourse Y which gives the discrete fuzzy centroid can be reduced to

$$y = [\sum_{i=1}^{I} y_i \mu_B(y_i)] / [\sum_{i=1}^{I} \mu_B(y_i)] \tag{6.2}$$

6.2.4 Tuning of FLC parameters

Previous experience with the controlled system is helpful in selecting the initial value of the FLC parameters. If insufficient information is available about the controlled system, FLC parameters can become a tedious trial-and-error process. The objective of the offline tuning algorithms used here is to change the controller gains to obtain the desired system response. The tuning algorithm tries to minimise three system performance indices. These indices are the system overshoot and the performance indices J1, J2 where:

J1 = ΣE^2
J2 = Σtime $\times E^2$ $\tag{6.3}$

where E is the system output error.

6.3 Example Simulations

A power system model consisting of a synchronous machine connected to a constant voltage bus through a double circuit transmission line is used in the simulation studies. A schematic diagram for the model is shown in Figure 6.2. The control signal generated by the proposed FLPSS is injected as a supplementary stabilising signal to the AVR summing point as shown in Figure 6.2.

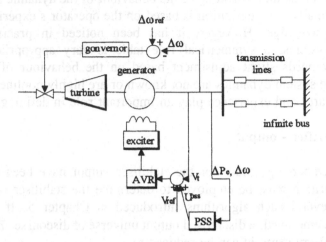

Figure 6.2 Schematic diagram of the power system model configuration

A number of studies have been reported in reference [1] and the results of these are compared with a conventional PSS (CPSS). For example, Figure 6.3 shows the response without stabiliser, and with FLPSS and CPSS under the following conditions: with the generator operating at a power of 0.9pu, 0.9 power factor lag, a 0.25 pu step decrease in input torque reference was applied at 2s, and the disturbance was removed at 9s and the system returned to the original operating point. The system without a stabiliser is highly oscillatory. Although the CPSS is effective in damping the oscillations, the load angle settles to its new value very smoothly and quickly with the FLPSS. The control generated by the two controllers is shown in Figure 6.4.

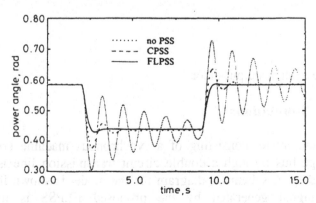

Figure 6.3 Response to ±0.25pu input torque disturbance at power=0.95pu and power factor=0.9lag

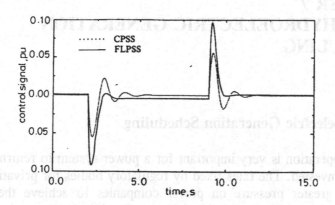

Figure 6.4 Stabilizing signal for CPSS and FLPSS

Experiments with a micro-computer based FLPSS as reported in reference [5] were performed at Itsukigawa Hydro-power station of Kyushu Electric Power Co. Ltd in 1992 and the filed test demonstrated its effectiveness.

References

1. K A El-Metwally, O P Malik, Fuzzy logic power system stabiliser, IEE Proc.-Gener. Transm. Distrib., Vol.142, No.3, 1995, pp.277 - 281
2. Y Y Hsu, C H Cheng, Design of fuzzy power system stabilisers for multimachine power systems, IEE Proc, Vol.137, Pt.C, 1990, pp.233 - 238
3. Y H Song, R K Aggarwal, A T Johns, Fuzzy logic protection for power systems, Electricity, No.3, 1994, pp.39 - 40
4. Y H Song, Novel adaptive control scheme for improving power system stability, IEE Proc, Vol.139, Pt.C, 1992, pp.423 - 426
5. T Hiyama, S Oniki, H Nagashima, Experimental studies on micro-computer based fuzzy logic power system stabilizer, Proc ANNPS'93, 1993, pp.212 - 217

CHAPTER 7
FUZZY HYDROELECTRIC GENERATION SCHEDULING

7.1 Hydroelectric Generation Scheduling

Economic operation is very important for a power system to return a profit on the capital invested. The rates fixed by regulatory bodies for privatised utilities place ever greater pressure on power companies to achieve the maximum possible efficiency. Operational economics involving power generation and delivery can be subdivided into two parts - one dealing with minimum cost of power production called economic dispatch and the other dealing with minimum-loss delivery of the generated power to the loads. For any specified load condition, economic dispatch determines the power output of each plant which will minimize the overall cost of fuel needed to serve the system load. Thus, economic dispatch focuses upon coordinating the production costs at all power plants operating on the system.

The major objective of hydrogeneration scheduling in a power system is to minimise the total fuel cost of thermal units by utilising the limited water resources. It is a typical optimisation problem in which the total operating cost over the study period is minimised subject to load and system constraints. The problem can be formulated as:

minimise

$$C = \sum_{t=1}^{24} COST_t \, (GTHERMAL_t) \qquad (7.1)$$

subject to

(i) The generation-load balance equations

$$GTHERMAL_t + \sum P_i \, (X_{it}) = L_t \qquad (7.2)$$

(ii) The water balance equations

$$Y_{it+1} = Y_{it} + \sum X_{jt} - X_{it} + \sum S_{it} - S_{it} + R_{it} \qquad (7.3)$$

(iii) Bounds on water releases

$$\underline{X_i} \leq X_{it} \leq \overline{X_i} \text{ and } \underline{S_i} \leq S_{it} \leq \overline{S_i} \qquad (7.4)$$

(iv) Bounds on a reservoir storage

$$\underline{Y_i} \leq Y_{it} \leq \overline{Y_i} \qquad (7.5)$$

where

C = system generation cost over study period
COST$_t$(.) = generation cost function at hour t which is approximated by a second-order polynomial
S$_{it}$ = spillage from reservoir during hour t

Since the convex objective function in eqn (7.1) is a nonlinear function of the variable GTHERMAL, it is necessary to linearise this function first in the linear programming formulation. Figure 7.1 depicts a piecewise linear approximation to the cost curve which is originally represented by a second-order polynomial.

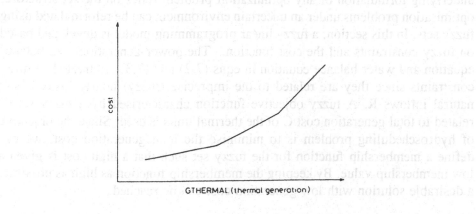

Figure 7.1 Thermal generation cost function approximated by multi-segment piecewise linear curve

The hydroscheduling problem can now be formulated as a general linear programming (LP) problem.

$$\min C = \underline{C}'\underline{X} \tag{7.6}$$

subject to

$$A\underline{X} \geq \underline{b} \tag{7.7}$$
$$\underline{X} \geq \underline{0} \tag{7.8}$$

where

C = total cost to be minimised
$\underline{X}' = \{X_1, \ldots, X_n\}$ = vector of control variables
$\underline{C}' = [c1, \ldots, cn]$ = vector of cost parameters
$\underline{b}' = [b1, \ldots, bm]$ = constraint vector
$A = [a_{ij}]$ = constraint matrix

Numerous approaches have been proposed for solving hydroelectric generation scheduling problem, including linear programming and dynamic programming.

7.2 Fuzzy Linear Programming Approach to Hydroelectric Generation Scheduling

7.2.1 Fuzzy linear programming

Fuzzy set theory is a generalisation of traditional crisp set theory. As the underlying formulation of any optimization problem relies on the set structure, optimisation problems under an uncertain environment can be reformulated using fuzzy sets. In this section, a fuzzy linear programming model is developed based on fuzzy constraints and the cost function. The power generation-load balance equation and water balance equation in eqns (7.2) and (7.3) are treated as fuzzy constraints since they are related to the imprecise (fuzzy) hourly loads L and natural inflows R. A fuzzy objective function characterised by a fuzzy set C_f related to total generation cost C of the thermal units is used. Since the objective of hydroscheduling problem is to minimise the total generation cost, we can define a membership function for the fuzzy set such that a high cost is given a low membership value. By keeping the membership function as high as possible, a desirable solution with low generation cost can be reached.

The hydroschduling problem originally formulated in eqns. 7.6-7.8 under crisp conditions must now be reformulated under fuzzy environments.

$$\min \; C_f \qquad\qquad (7.9)$$

subject to
$$A\underline{X} \geq \underline{b} \qquad\qquad (7.10)$$
$$\underline{X} \geq 0 \qquad\qquad (7.11)$$

where C_f in eqn (7.9) is a fuzzy objective and eqn. (7.10) is a fuzzy constraint.

In the proposed approach, the fuzzy objective in eqn. (7.9) is reformulated as a fuzzy constraint. In other words, we want to keep the total cost below some maximum level of expense, C^M. Below this level of expense, a higher membership value indicates a better solution for that objective. Thus eqns (7.9) - (7.11) are formulated as follows. Find $\underline{X}$ such that

$$B\underline{X} \geq \underline{d} \qquad\qquad (7.12)$$
$$\underline{X} \geq 0 \qquad\qquad (7.13)$$

where

$$B^t = (-C^t \; A) \qquad \underline{d}^t = (-C^M \; \underline{b}) \qquad\qquad (7.14)$$

Note the C^M indicates the worst cost acceptable for the objective function which is determined by the operators according to their experience.

There are three types of fuzzy constraints in eqn. (7.12), which will be discussed later. Each constraint defines a fuzzy region of acceptability. The intersection of all such regions is the fuzzy set of feasible solutions.

With the membership functions for the three types of fuzzy constraints which will be detailed later, we can proceed to determine an "optimal" solution which best satisfies these constraints. Based on previous discussions, a solution to the fuzzy linear programming model is called a feasible solution if X satisfies the crisp constraints and the bounds for hourly loads and natural inflows are satisfied. Among the numerous feasible solutions to the FLP problem, we will choose a solution which meets the fuzzy constraints to the highest degree.

It was mentioned that the fuzzy constraints in eqn. (7.12) can be classified into three types and we can use membership functions $\mu_C(X)$, $\mu_{Lt}(X)$, $\mu_{Rit}(X)$ to describe the degrees to which the constraints related to generation cost, generation-load balance and water balance, respectively, are satisfied. Each constraint defines a fuzzy region of acceptability. The intersection of all such regions is the fuzzy set of feasible solutions.

Assume that the minimum operator is to be used for fuzzy set intersection. The constraints can be written as

$$S = \min \, [\mu_C(X), \mu_{Lt}(X), \, \mu_{Rit}(X)] \tag{7.15}$$

where $\mu_C(X)$, $\mu_{Lt}(X)$, $\mu_{Rit}(X)$ are the fuzzy sets for the three types. The higher the membership values the better the solution. Thus, the best solution is:

$$S^* = \max\text{-}\min \, [\mu_C(X), \mu_{Lt}(X), \, \mu_{Rit}(X)] \tag{7.16}$$

7.2.2 Fuzzy constraint related to objective function

It was mentioned before that the generation cost should be less than CM. In addition, a low cost must be given a high membership value. Figure 7.2 depicts the membership function for the fuzzy variable signifying total cost C. In this figure, a membership or a satisfaction value of 1 is assigned to any C that is less than C^M-Pc. As C becomes larger than C^M-Pc, the degree of satisfaction will decrease to zero linearly when C is equal to C^M. The degree of satisfaction is zero for any value of C greater than C^M. Note that a linear membership function is adopted since we are using the fuzzy LP approach.

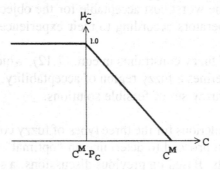

Figure 7.2 Membership function for fuzzy generation cost C

7.2.3 Fuzzy constraints related to generation-load balance equations

It is noted from the generation-load balance equations in eqn. 4 that the sum of the generation from thermal units, CTHERMAL, and the total hydrogenerations $\sum P_i(X_{it})$ must be equal to the hourly load demand L_t. In the hydroscheduling problem, L_t denotes the hourly load demands in the future. Therefore, L_t can only be reached through load forecasting and there are always errors in the forecast hourly loads. As result, the actual load, Ltactual, can be expressed as the sum of the forecast load, $L_{tforecast}$, and the forecast error ΔL_t. Note that the forecast load $L_{forecast}$ is crisp while the forecast error and the actual load are imprecise and are characterised by fuzzy sets. As long as the membership function for the fuzzy set ΔL_{if} is known, that for the fuzzy set L_f can be determined. Here, a linearly decreasing membership function in a triangular form is used. Thus a membership function for load demand shown in Figure 7.3 can be employed.

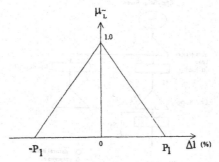

Figure 7.3 Membership function for load demand

7.2.4 Fuzzy constraints related to water balance

From eqn. 7.3, the water balance equation for reservoir i can be written as

$$Y_{it+1} - Y_{it} - \sum X_{jt} + X_{it} - \sum S_{lt} + S_{it} = R_{it} \qquad (7.17)$$

Just as in the case of hourly load demands, there are also errors in the forecast natural inflows R_{it}. Thus, the actual natural inflow R_{actual}, is the sum of the forecast natural inflow, $R_{forecast}$, and the forecast error, ΔR. Based on the previous experience, a membership function μ_R as shown in Figure 7.4 is employed for the fuzzy set R_f.

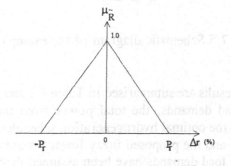

Figure 7.4 Membership function μ_R for natural inflow

7.3 Example Simulations

A practical system [1] was employed to demonstrate the effectiveness of the proposed FLP based hydroelectric generation scheduling. The schematic diagram is shown in Figure 7.5.

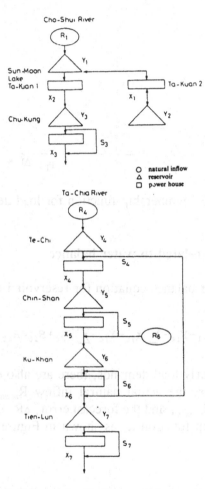

Figure 7.5 Schematic diagram of the example system

The computational results are summarised in Table 7.1 and Figure 7.6. Table 7.1 gives the hourly load demands, the total power from the hydrounits and the production costs for the optimal hydrogeneration schedules reached by using the linear programming and the proposed fuzzy linear programming. It is observed that different hourly load demands have been assumed ih the two approaches. In the LP approach, the forecast hourly loads are directly employed. But these load demands have been modified in the FLP approach by taking the uncertainties in the load demands into account.

Table 7.1 Results from LP and FLP

Hour h	Load demand, MW		Hydroelectric generation, MW	
	LP	FLP	LP	FLP
1	7624	7620	−100	−97
2	7340	7340	−678	−678
3	7097	7096	−693	−687
4	6926	6925	−750	−750
5	6818	6818	−750	−750
6	6762	6761	−750	−750
7	6916	6916	−750	−750
8	7467	7467	−119	−119
9	9461	9431	323	393
10	1019	910144	789	789
11	10531	10475	1469	1468
12	10620	10562	1851	1851
13	9567	9541	877	877
14	10744	10683	1851	1851
15	10982	10922	1851	1851
16	10823	10766	1851	1851
17	10560	10503	1850	1851
18	9726	9689	756	759
19	9683	9648	630	677
20	9897	9850	327	334
21	9690	9648	264	275
22	9295	9263	203	204
23	8968	8944	11	22
24	8476	8465	−710	−710

α = 0.951727
Total cost from LP = 123 163 560 NT$ = 1.0 pu
Total cost from FLP = 122 368 292 NT$ = 0.993543 pu

Figure 7.6 illustrates the water release schedules for two of the ten reservoirs over the 24h scheduling period. It is observed that different water release schedules have been obtained by using the two approaches since different hourly inflows are assumed by the two approaches. In the LP approach, the forecast hourly inflows are directly employed. But these natural inflows have been modified in the FLP approach by taking the uncertainties in the natural inflows into account.

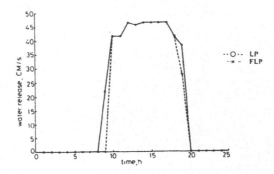

Figure 7.6 Water release from Ta-Kuan hydroplant from LP and FLP

References

1. R H Liang, Y Y Hsu, Fuzzy linear programming: an application to hydroelectric generation scheduling, IEE Proc. Gener.Transm.Distrib., Vol.141, No.6, 1994, pp.258 - 574
2. K Tomsovic, A fuzzy linear programming approach to the reactive power/voltage control problem, IEEE Trans Power Systems, Vol.7, No.1, 1992, pp.287 - 293
3. K H Abdul-Rahman, S M Shahidephpour, A fuzzy-based optimal reactive power control, IEEE Trans Power Systems, Vol.8, No.2, 1993, pp.662 - 670
4. D Srinivasan, C S Chang, A C Liew, Multiobjective generation scheduling using fuzzy optimal search technique, IEE Proc DTD, Vol.141, No.3, 1994, pp.233 - 242

CHAPTER 8
NEURAL NETWORKS: AN INTRODUCTION

8.1 Artificial Neural Networks

8.1.1 Artificial neurons

A Neural Network (NN) is an implementation of an algorithm inspired by research into the human brain. It is obvious that the human brain is superior to a digital computer at many tasks. A good example is the processing of visual information; for example, a one-year old baby is much better and faster at recognising objects, faces, etc. than even the most advanced Artificial Intelligence (AI) system running on the fastest super-computer system. The brain has many other attributes that are desirable in AI systems:

- it is robust and fault tolerant; nerve cells in the brain die every day without significantly affecting its performance

- it can deal with information that is fuzzy, probabilistic, noisy or inconsistent

- it is highly parallel

- it is small, compact and dissipates very little power

Only in tasks based primarily on simple arithmetic does the computer outperform the brain. This is the real motivation for studying neural computing. It is an alternative computational concept to the conventional approach which is based on a programmed instruction sequence. It is inspired by knowledge from neuroscience, though it does not try to be biologically realistic in detail. Its practical applications lie mainly in computer science and engineering.

The field is also known as neuro-computation, collective computation, connectionism, etc. The term neural network implies that it was originally aimed more towards modelling networks of real neurons in the brain. However, the models are extremely simplified compared to the latter, though they are very valuable for gaining an insight into the principles of biological computation.

Neurons: The brain is composed of about 10^{11} neurons (nerve cells) of many different types. Figure 8.1 is a schematic of a single neuron. As can be seen, tree-like networks of nerve fibre called dendrites are connected to the cell body where the cell nucleus is located. Extending from the cell body is a single long fibre called the axon which eventually branches into strands or sub-strands. At the ends of these are the transmitting ends of the synaptic junctions (or synapses) to other neurons. The receiving ends of these junctions on other cells can be found both on the dendrites and on the cell bodies themselves. The axon of a typical neuron makes a few thousand synapses with other neurons.

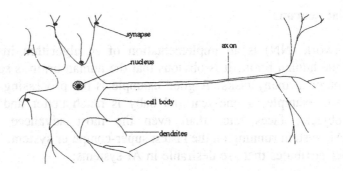

Figure 8.1 Schematic diagram of a typical neuron

The transmission of a signal from one cell to another at a synapse is a highly complex process in which specified transmitter substances are released from the sending side of the junction. The effect is to raise or lower the electrical potential inside the body of the receiving cell. If this potential reaches a threshold, a pulse or action potential of fixed strength and duration is sent down the axon. We then say that the cell has fired. After firing, the cell has to wait for a time called the refractory period before it can fire again.

8.1.2   Characteristics of artificial neural networks (ANNs)

An ANN consists of a number of neurons (or nodes) which are the elementary processing units that are connected together according to some pattern of connectivity. Figure 8.2 shows a simple model of a neuron as a binary threshold unit. Neuron j is characterised by the number of inputs X_1, X_2,---X_n, the weights W_{j1}, W_{j2},---W_{jn} connecting each input to the neuron, its activation function F and its output Y_j; n is the number of inputs to a neuron.

The inputs of a particular type are combined together to give a net input n_j of the jth unit. For example, if there are only excitatory (or positive) inputs to a particular neuron, the net input is simply the weighted sum of the inputs into the

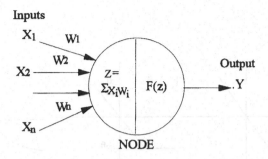

Figure 8.2 A neural network node

neuron as given by:

$$n_j = \sum_{k=1}^{n} W_{jk} X_k \qquad\qquad (8.1)$$

As shown above, ANNs gain their overall processing capability by connecting these simple neurons to other neurons with an associated weight W. This weight determines the structure of the signal which is transmitted from one neuron to another. The total collection of weights are the parameters that completely specify the model of the process which this network represents.

The neuron uses this net input, together with information on its current activation state to determine its new state of activation, ie the output Y_j, given as:

$$Y_j = F(n_j) = F[\sum_{k=1}^{n} W_{jk} X_k] \qquad\qquad (8.2)$$

There are several activation functions in use. For example, in one case Y_j is either 1 or 0 and represents the state of neuron j as firing or not firing respectively. In this case F(z) is the unit step function given as:

$$F(z) = \begin{cases} 1 \text{ if } z \geq 0; \\ 0 \text{ otherwise} \end{cases} \qquad\qquad (8.3)$$

There are a number of other activation functions in use. The most frequently used ones are the identity, the linear threshold and the sigmoid function as shown in Figure 8.3. It should be mentioned that the sigmoid function is by far the most common form of activation function used in ANNs. It is defined as a strictly increasing function that exhibits smoothness and asymptotic properties. A sigmoid can be mathematically represented by the hyperbolic tangent function defined by $F(z) = Tanh\ z$.

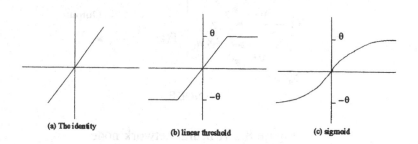

(a) The identity   (b) linear threshold   (c) sigmoid

Figure 8.3 Activation functions

The neurons are normally connected to each other in a specified fashion to form the ANN. These arrangements of interconnections form a single layer or several layers. In a large number of ANN models, such as the Perceptron, Linear Association and Multi-layer Feed Forward networks (Figure 8.4 shows a typical example of the latter), the output from the units from one layer is only allowed to activate neurons in the next adjacent layer. However, in some models such as the Kohonen networks, the signal is allowed to activate neurons in the same layer. The weights have the biological interpretation of the synaptic interconnection strength between neurons. Figure 8.5 shows a general ANN architecture. Different types of ANNs will be considered in some detail in the next section.

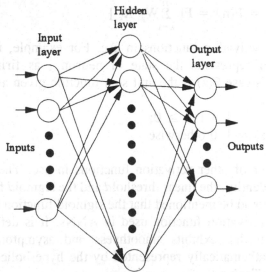

Figure 8.4 General feed forward network architecture

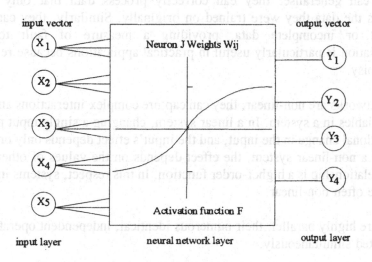

input vector   Neuron J Weights Wij   output vector

Activation function F

input layer   neural network layer   output layer

Figure 8.5 General neural network architecture

Lastly, the ANN must have a mechanism for learning (also called training). Learning is done for a subset of input vectors, called a training set, whose properties are known or representative. Learning alters the weights associated with the various interconnections and thus leads to a modification in the strength of interconnections.

Thus an ANN is characterised by its architecture, its processing algorithm and its learning algorithm. The architecture specifies the way the neurons are connected. The processing algorithm specifies how the ANN with a given set of weights calculates the output vector Y for any input vector X. The learning algorithm specifies how the ANN adapts its weights for all given training vectors X.

Before we consider a few different classes of ANNs that could be of interest for power systems, there are a number of issues to be considered.

(i) Advantages of ANNs

1) They are adaptive: they take data and learn from it. Thus they infer solutions from the data presented to them, often capturing quite subtle relationships. This ability differs radically from standard software techniques because it does not depend on the programmer's prior knowledge of rules. NNs can reduce development time by learning underlying relationships even if they are difficult

to find and describe. They can also solve problems that lack existing solutions.

2) NNs can generalise: they can correctly process data that only broadly resembles the data they were trained on originally. Similarly, they can handle imperfect or incomplete data, providing a measure of fault tolerance. Generalisation is particularly useful in practical applications because real world data is noisy.

3) The networks are non-linear: they can capture complex interactions among the input variables in a system. In a linear system, changing a single input produces a proportional change in the input, and the input's effect depends only on its own value. In a non-linear system, the effect depends on the values of other inputs, and the relationship is a higher-order function. In this respect, systems in the real world are often non-linear.

4) NNs are highly parallel: their numerous identical, independent operations can be executed simultaneously.

(ii) Disadvantages of ANNs

1) There is no definite way of choosing the optimum architecture.

2) There is no definite way of finding the best solution.

3) The solution depends upon the accuracy of the training set

(iii) Other issues

Massive parallelism in computational networks is extremely attractive in principle, but in practice, there are many other issues to be decided before a successful implementation can be achieved for a given problem:

- What is the best architecture? Should the units be divided into layers or not? How many connections should be made between units and how should this be organised? What sort of activation function F(z) should be used?

- How can a network be programmed? Can it learn a task or must it be predesigned? If it can learn a task, how many examples are needed for good performance? How many times must it go through the examples? Does it need the right answers during training, or can it learn from correct/incorrect reinforcement? Can it learn in real-time while functioning or must the training phase be separated from the performance phase?

- What can the various types of network do? How many different tasks can they learn? How well? How fast? How robust are they to missing information, incorrect data and unit removal or malfunction? Can they *generalise* from known tasks or examples to unknown ones? What classes of input-to-output functions can they represent?

- How can a network be built in hardware? What are the advantages and disadvantages of different hardware implementations and how do they compare to simulations in software?

These questions are obviously interfaced and cannot be answered independently. Although a number of them will be answered when we consider different classes of ANNs, three of the issues - namely hardware, generalisation and programming - need to be elaborated upon.

Hardware: Almost everything in the field of neural computation has been done by simulating the networks on serial computers (more recently on parallel processors) or by theoretical analysis. Neural network VLSI (very large-scale integration) chips are far behind the models. The main problem with ANN chips is that one needs many connections, often some fraction of the square of the number of units. The space taken up by the connections is usually the limiting factor for the size of a network and the neural chips made hitherto contain far too few units for most practical applications.

Practical alternatives to integrated circuit chips include optical computers. However, this field is still very young and much research needs to be done in this area.

Generalisation: The reason for much of the excitement about ANNs is their ability to generalise to new situations. After being trained on a number of examples of a relationship, they can often induce a complete relationship that interpolates and extrapolates from the examples in a sensible way, but what is meant by sensible generalisation is often not clear. In many problems, there are almost infinitely many possible generalisations. How does a NN choose the *right* one?

Programming: In the case of ANNs, the concept of programming is radically different from that understood and as applied in the case of conventional computing, in the sense that we *teach* the NN to perform the desired computation by iterative adjustments of the W_{jk} weights (or strengths). This may be divided into two main ways - Supervised and Unsupervised learning:

In supervised learning, each example consists of a set of initial conditions and a set of the resulting actions or decisions. This situation is equivalent to having a teacher who provides the correct answers from which the system can learn. Figure 8.6 illustrates the process of supervised learning.

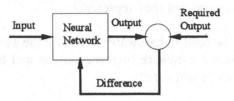

1. Present an input to the Net
2. Calculate the difference between the actual and the required output
3. Use this to alter the weights in the Net
4. Repeat for another input

Figure 8.6 Training of a supervised neural network

In contrast, each example used by unsupervised learning methods consists of the initial conditions only ie, there are no appropriate actions or decisions associated with the initial conditions. Here, the aim is to understand the underlying statistical structure or patterns in the area.

In supervised learning, one thus provides the input attribute vector and the associated target vector, while in unsupervised learning, one provides only the input attribute vector.

Next we consider a few different classes of ANNs that are of interest for power systems.

The architecture specifies the way the neurons are interconnected and the actual method of determining the output for a given set of inputs is called the processing algorithm. Different NNs are characterised by differences in the architecture, the learning algorithm and the processing algorithm.

In using NNs, the first step is learning the weights to model the process. This is called the learning phase. Once these weights have been learnt, when an input is used to stimulate the network, it is used to determine the output. This is called the prediction or recall phase. One of the big advantages of the NN is that the recall phase is very fast.

Next we consider the different classes of NNs widely used in power systems.

8.2 Neural Network Types

8.2.1 Single layer linear models (supervised)

Single-layer ANNs consist of a set of N input units and a set of n output units. Each output unit may be interconnected with all the input units as illustrated in Figure 8.7.

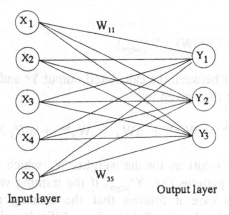

Figure 8.7 Single layer perceptron architecture

The strength of the inter-connectivity can be represented as a weight matrix W with positive (excitatory), negative (inhibitory) or zero (no connection) values. The activation function F is the identity (Figure 8.3). Thus:

$$Y_j = \sum_{k=1}^{n} W_{jk} X_k \quad \text{for } j = 1, ..., n \qquad (8.4)$$

The weights can be learnt by training the network using a training set of M target vectors Y^μ_{target} for the output for a given set of M input vectors X^μ where $1 \le \mu \le M$. The common approach to training the network to learn the appropriate weights is based on the Delta Rule. This rule adjusts the weights during training such that the mean square error between the actual vector Y^μ obtained for the training input vector and the known target Y^μ_{target} is minimised for one training pattern (X^μ, Y^μ_{target}) at a time. Thus the weight matrix W is calculated such that:

$$E(W) = \sum_{j=1}^{n} [Y^\mu_j(W) - Y^\mu_{j \ target}]^2 \qquad (8.5)$$

is minimised. The steepest descent rule is then used to find the minimum of E using an iterative adjustment step to the weights. It should be noted that the training procedure for layered perceptrons based on the Delta rule or the error back propagation is called a steepest descent method for finding the minimum of a function. In this technique, at the current point in the weight space, we compute the steepest slope and take a step in that direction thereby changing our location in weight space. The process is repeated until an acceptably low error is obtained. We thus obtain the weights of the form:

$$\Delta W_{jk} = -\eta\ \partial E/\partial W_{jk} = \eta\ \delta^\mu_j\ X^\mu_k\ ,\ j = 1,\ ...,\ n;\ \ k=1,\ ...,\ N \tag{8.6}$$

Where η $(0 < \eta < 1)$ is a (usually decaying) search parameter called the learning rate and:

$$\delta^\mu_j = (Y^\mu_j(W) - Y^\mu_{j\ target}) \tag{8.7}$$

This is the delta error between the calculated output Y^μ and target vector Y^μ_{target}. The new weights are then given by:

$$W_{jk}^{new} = W_{jk}^{old} + \Delta W_{jk} = W_{jk}^{old} + \eta\ \delta^\mu_j\ X^\mu_k \tag{8.8}$$

The delta rule yields solutions for the weights W_{jk} which minimise E(W) for a given set of training patterns (X^μ, Y^μ_{target}) if the training vectors X^μ are linearly independent. In this case it follows that the training patterns are linearly separable. However, the above techniques have difficulty with training sets where a non-linear relationship exists between the inputs and outputs.

8.2.2 Multi-layer feed-forward networks with error back-propagation learning rule (supervised learning)

A class of ANNs that overcome the aforementioned limitations of single layered linear networks and which are able to model non-linear relationships between the inputs and outputs, is the multi-layer feed-forward network with learning carried out using the error back-propagation rule. This model is commonly referred to as the Multi-Layer Perceptron (MLP).

This network consists of a set of N input units, a set of n output units and one or more layers of J intermediate units. These intermediate layers are called *hidden* layers since the units in them do not directly communicate with the environment. A multi-layer network with one hidden layer is shown in Figure 8.8. Hidden units V_j, $j = 1,\text{---}J$ and output units Y_i, $i = 1,\text{----}n$ process their inputs as in the case of the single layer ANN. Thus:

$$V_j = F[\ \sum_{k=1}^{n} w_{jk}\ X_k\ ]\quad \text{where } j = 1,\ ...,\ J \tag{8.9}$$

$$Y_i = F[\ \sum_{m=1}^{J} W_{im}\ V_m\ ]\quad \text{where } i = 1,\ ...,\ n \tag{8.10}$$

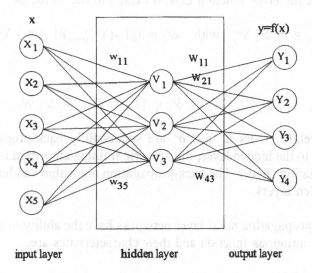

Figure 8.8 Neural network architecture of the multi-layer perceptron with 1 hidden layer

The major step forward in the use of hidden units was the development of a method of learning that would alter the weight matrices Δw and ΔW. This learning rule is a generalisation of the Delta Rule for multi-layer networks. It carries out a minimisation of the mean square error E between the obtained outputs corresponding to the input states and the desired target states, as for the single layer network. Note that here the error E is a function of both weight matrices w and W because the output Y_i depends on the weights W_{ij} and the output of the hidden unit V_j, the latter depending on w_{jk}. We thus have:

$$E(w,W) = \sum_{i=1}^{n} (Y^{\mu}_i(w,W) - Y^{\mu}_{i\ target})^2 \qquad (8.11)$$

The hidden units carry out a recoding of the input patterns X^{μ} which are then fed into the output units to permit an arbitrary mapping of inputs to outputs. Both the hidden units and output units will be assumed to have an activation function which is a sigmoid F as shown in Figure 8.3.

The algorithm carries out a steepest descent correction on the matrix giving Δw and ΔW. The difference between the MLP and the single layer network is that both Δw and ΔW are adjusted as shown by:

$$\Delta W_{ij} = -\eta\ \partial E/\ \partial W_{ij} \quad \text{and} \quad \Delta w_{jk} = -\eta\ \partial E/\cdot\partial w_{jk} \qquad (8.12)$$

For the non-linear activation function $F(z) = \text{Tanh } z$, a straightforward derivation of the error function $E(w,W)$ leads to the formulas:

$$\Delta W_{ij} = \eta \, \Delta^\mu_i \, V^\mu_j \quad \text{with} \quad \Delta^\mu_i = ([1 - (Y^\mu_{i \text{ target}})^2] \, [Y^\mu_i - Y^\mu_{i \text{ target}}] \qquad (8.13)$$

and

$$\Delta w_{jk} = \eta \, \delta^\mu_j \, X^\mu_k \quad \text{with} \quad \delta^\mu_j = [1 - (V^\mu_j)^2] \sum_{i=1}^{n} \Delta^\mu_i \, W_{ij} \qquad (8.14)$$

Since the weight errors Δ^μ_i and δ^μ_j are successfully back-propagated from the output layer to the hidden layer, this specific training algorithm is known as error back-propagation (EBP). The back-propagation algorithm can be generalised for several hidden layers.

These back-propagation multi-layer networks have the ability to approximate any non-linear continuous function and their characteristics are:

- feed forward processing
- supervised learning using error back propagation
- multi-layer architecture
- approximation of a non-linear continuous or discrete function

However, they may present a number of difficulties and these include:

- they frequently have a local minimum at which the solution processes can get stuck
- their convergence is slow in the presence of 'valleys' because of weaknesses in the steepest descent technique. This can be overcome partially by using momentum terms for acceleration or by using conjugate gradient approaches

- choice of learning parameters can be difficult

- large systems could have impractically long convergence times.

MLPs are used in nearly every area of power systems whose tasks can be formulated as an approximation problem. For example, in the area of load forecasting, the MLP is used for the prediction of the hourly load as a function of the previous 24 hours (this will be discussed later in some detail).

8.2.3 Kohonen network (unsupervised learning)

The Kohonen network, also known as the self-organising feature map (SOFM), is an example of unsupervised learning. The purpose of SOFM is that patterns of high dimension (ie long vectors) are transformed into one or more dimensional patterns.

The SOFM has two purposes:

1) the clustering of the input space into a finite number of classes represented by the NN's weight vectors,

2) the topology-preserving mapping of high dimensional input vectors (ie long vectors) onto a lower dimensional surface represented by the location of the neuron on the grid.

The Kohonen network has the ability to find clusters in the data as well as structure and to perform an ordered or topology-preserving mapping, thus revealing existing similarities in the inputs. The topology preserved with this network need not correspond to a physical arrangement; it can correspond to any statistical feature of the input set.

(1) The self-organising feature map

Self-organising feature maps are NNs which map N-dimensional vectors to a two-dimensional NN in a non-linear fashion, preserving a topological order, ie input data representing similar characteristics are mapped to contiguous clusters. We define that two vectors are similar if their Euclidean distance is small. The network learns without supervision; this means that the topological order of the input data is not necessarily known previously.

The self-organising feature map consists of M neurons, usually arranged on a square grid and connected according to a given neighbourhood relation. Each neuron possesses a weight vector of dimension N, which is also the dimension of the input vectors that are to be classified. The components of the weight vector w_k can be represented as connections between each component of the input vector and neuron k. In this case, the dimension N of the input vectors equals 2L.

Fig 8.9 shows a Kohonen network which maps three-dimensional input vectors on a two-dimensional map containing 16 neurons. Each component j of the input vector is connected to every neuron i of the map by a weight w_{ij}. Only the weights belonging to neurons 0 and 8 are represented in Figure 8.9. A neuron in this square network has direct neighbours, called neighbours of first order, except for the neurons along the borders of the grid. Two neurons i and j are

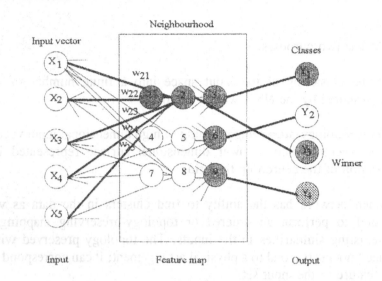

Figure 8.9 Structure of the self-organising feature map

(2) The self-organising algorithm

Let M be the number of neurons with weight vectors $w_k \epsilon R^N$, $1 \le k \le M$ and $X = \{x \epsilon R^N \mid x \text{ training vector}\}$, the training set. At each step t of the training phase, a vector x_t drawn randomly from the training set X, is presented as input to the network. The neuron most receptive to this input vector, ie the neuron c whose weight vector is the closest to the input vector in the sense of the Euclidean distance, is selected. The weight vector w_c of this neuron is then adapted, becoming closer to the input vector. The neighbouring weight vectors are also adapted. The parameters of the adaptation rule are such that the weight vectors converge to an equilibrium after several thousand input vectors have been presented.

The remarkable property of the Kohonen network is that this NN learns the topology of the N-dimensional vector space with a given set of training vectors. By learning the topology, we mean that vectors that are neighbours in the N-dimensional space of the input vectors will also be neighbours in the two-dimensional space of the Kohonen feature map. The set of input vectors, which may amount to several thousand vectors of N dimensions, will be represented after learning by a smaller number M of N-dimensional weight vectors.

(i) Unsupervised learning

The components of all weight vectors are randomly initialised. The learning step t is initialised to 0. The following procedure implements the unsupervised self-organisation:

1 Choose input vector $x \epsilon X$ randomly in the training set.

2 Select the neuron c with the weight vector closest to the training vector:

$$\| w_c(t) - x \| = \min \| w_k(t) - x \| \quad \text{for all k} \tag{8.15}$$

3 Define a neighbourhood around the selected neuron as a decr   .ng function of the steps of the learning process.

The neighbourhood order function is defined as:

$$k(t) = [k_o \exp(-t/\tau_k)] \tag{8.16}$$

where [x] denotes the largest integer less than or equal to x. The neighbourhood set $N_d(t,c)$ of neuron c is determined such that neuron $i \epsilon N_d(t,c)$ if i is a neighbour of c of order d with $d \leq k(t)$.

4 Update the weight vector w_c of the selected neuron c and the weight vectors w_k of its neighbours according to the rule:

$$w_k(t) = w_k(t - 1) + \alpha(t) [x - w_k(t - 1)], \text{ if neuron k belongs to the neighbourhood} \tag{8.17}$$

$$w_k(t) = w_k(t - 1) \quad \text{otherwise}$$

5 Increment the time $t = t + 1$

6 If $\| w_k(t) - w_k(t - 1) \| > \varepsilon$ goto 1 else stop

The neighbourhood function, represented in Figure 8.10, is defined as:

$$\alpha(t,i,c) = \begin{cases} (\alpha_o \exp(-t/\tau_\alpha))/(1 + d) & \text{for } i \epsilon N_d(t,c) \\ 0 & \text{elsewhere} \end{cases} \tag{8.18}$$

Figure 8.10 illustrates how the neighbourhood function $\alpha(t,i,c)$ decreases in time. In the beginning, the nearest weight vector chosen in step 2 will be modified with a weaker factor $\alpha(t,i,c)$. For large t, $\alpha(t,i,c)$ becomes very small; the value of the input vector influences the modification of the weight vectors only slightly. The convergence criterion ε determines when $\alpha(t,i,c)$ is considered to be sufficiently small.

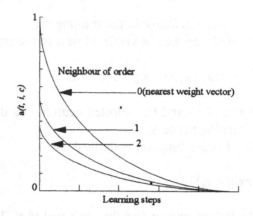

Figure 8.10 Example of neighbourhood function $\alpha(t,i,c)$ for $i = 0,1,2$

(ii) Classification

In the classification phase, the network maps a vector with unknown features to the cluster where its closest neighbours have been mapped to the algorithm:

1 Present the input vector x.
2 Select the neuron c with the weight vector closest to the input vector:

$$\| w_c(t_{max}) - x \| = \min \| w_k(t_{max}) - x \| \quad \text{for all k} \qquad (8.19)$$

The characteristics of the Kohonen network model are:

- winner-takes-all processing
- unsupervised learning
- lateral connected architecture

used for

- topology preserving clustering
- classification

As examples of applications in power systems, in the area of security assessment, SOFM is used in order to reduce the space of all feasible operating points into a finite set of typical operating points; in the area of load forecasting, the SOFM creates classes of load patterns which are averages of several similar load patterns such as Sundays in Summer or weekdays in spring.

8.2.4 Hopfield networks (supervised learning)

There are two types of Hopfield networks; the first is the discrete output (0-1) stochastic network and the second has a deterministic continuous form. Here we discuss the discrete version which is the one used in power system applications. The Hopfield net is used either as an associative memory or for optimisation.

Figure 8.11 shows the architecture of the Hopfield model, which is a fully connected recurrent NN consisting of N neurons. The weights w_{ij} are placed on the lateral connections between neuron i and j. The weight matrix is symmetric with zero diagonal terms.

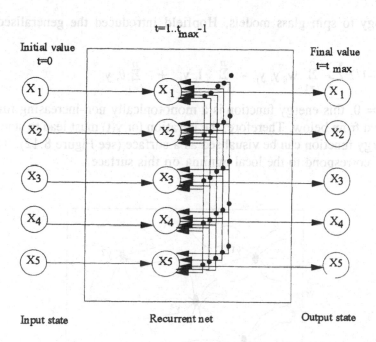

Figure 8.11 Neural network architecture of the Hopfield network

Each neuron has two states characterised by its output y_j which are y^0_j and y^1_j, usually -1 and 1. Input to a neuron can be:

    (i) An external input I_j called the bias term.
    (ii) Initial input x_j at step t = 0 and outputs from other neurons $y_i(t)$ for t > 0.

Let us assume for the moment that the weights are given. Then at step t for t > 0, the total input to neuron j is:

$$net_j(t) = \sum_{i=1}^{n} w_{ij}\, y_i(t-1) + I_i \,,\ t > 0 \qquad\qquad (8.20)$$

The motion of the state of the system with N neurons in state space describes the computation being performed. Any model must describe evolution of the state with time. Hopfield proposed a model with stochastic evolution. Each neuron samples its input at random. It then changes the value of its output (or leaves it unchanged) according to a threshold function g as follows:

$$y_j(t)\ g[\ \sum_{i=1}^{n} w_{ij}\, y_i(t-1) + I_i]\ \text{ with }\ g(y_j) = \left\{ \begin{array}{l} y^0_j\ \text{ for }\ y_j \le \theta_j \\[1em] y^1_j\ \text{ for }\ y_j > \theta_j \end{array} \right. \qquad (8.21)$$

In analogy to spin glass models, Hopfield introduced the generalised energy function:

$$E(Y) = -1/2 \sum \sum_{j=1}^{n} w_{ij}\, y_i\, y_j \ -\ \sum_{i=1}^{n} I_i\, y_i \ +\ \sum_{i=1}^{n} \theta_i\, y_i \qquad (8.22)$$

For $w_{ij} = 0$, this energy function is a monotonically non-increasing function of y bounded from below. Therefore, the iteration for y(t) must lead to stable states. The energy function can be visualised as a surface (see Figure 8.12). The stable states y^μ correspond to the local minima on this surface.

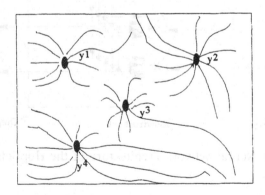

Figure 8.12 Energy surface with local minima

For a given set of fixed stable points, the model behaves as an associative memory (CAM-Content Addressable Memory). A stable fixed point has to be specified such that provided you started in a region of the state space near the point, you would converge to this nearest fixed stable point. We can think of the location of a particular fixed stable point in the state space as the information content of the memory. States near a given fixed stable point can be considered as having partial or noisy information on the memorised information. Moving an initial point with partial information to the fixed stable point containing the memorised information corresponds to a recall of information based on an association between the noisy information and the memorised 'ideal' information. Instead of supplying an address of the memorised information, some part of the content of the memory is used for its recall. Hence we have a CAM. In this case the bias term I_i and the threshold θ_i are zero for all i.

In the case of an auto-associative memory, the weights of the Hopfield network have to be determined such that the training vectors x^μ to be recognised correspond to these stable fixed points y^μ. This can be achieved by storing the weights as the outer product of all couples of training vectors:

$$w_{ij} = 1/N \sum_{\mu=1}^{M} x_i^\mu x_j^\mu \quad \text{for} \quad i = j \quad \text{and} \quad w_{ij} = 0 \qquad (8.23)$$

For an optimisation task with quadratic cost function, the parameters of the cost function to be optimised have to be matched to the terms w_{ij}, I_i and θ_i of the generalised energy function E(y). Thus the weights of w_{ij} of the Hopfield network are given as the factors of the second order terms of the input vector components of the cost function. The iteration process for y(t) will then converge to one of the local minima of the energy function which by construction is a local minimum of the cost function.

The convergence of the algorithm towards one of the local minima of this energy function as well as the results on the limited storage capacity can be derived by methods from statistical mechanics. The major drawback of the Hopfield network is the limited storage capacity of the associative memory. It has been shown that for N neurons, only approximately 14xN/100 patterns can be stored effectively. The energy function further has local minima (so called spurious states) in addition to those specified by the cost function or the memorised patterns respectively. These minima thus present a good but not necessarily optimal solution to the optimisation problem (nor do they represent any memorised pattern).

The characteristics of the Hopfield network are:

- iterative processing

- supervised learning with weights defined by

       - the pattern to be recognised

or

       - cost function to be minimised

- fully connected architecture

- content addressable associative memory

- optimiser of a quadratic energy function

The Hopfield network acts as an associative memory where noisy patterns converge to the idealised stored pattern. This effect can be used in the area of state estimation. In the power systems area, the hopfield network is mainly studied as an optimiser as, for example, for optimal load flow, generation expansion, unit commitment and economic dispatch.

8.3 Neural Networks in Power Systems

Modern power systems are required to generate and supply high quality electrical energy to consumers. In order to achieve this requirement, computers are extensively applied to power system operation, planning, monitoring and control; power system application programs for analysing system behaviour are stored in computers. For example, at the planning stage of a power system network, system analysis programs are executed repeatedly; engineers adjust and modify the input data to these programs according to their experience and heuristic knowledge about the system until satisfactory plans have been determined. However, the programs developed have hitherto been based on mathematical models and implemented using languages suitable for numeric computation only. For sophisticated approaches to system planning, for example, the development of methodologies and techniques to incorporate practical knowledge of power engineers into programs (which also include the numeric analysis programs) is needed.

Current approaches to power system computation can be broadly categorised into the following:

1) detailed models of a component with a simplified view of the system to give qualitative insights, for example, D-Q model of a generator,

2) mathematical models of the whole system as the basis for algorithmic solutions to the problems such as load flow and transient stability studies,

3) expert systems which represent an explicit but nor necessarily a formal model of the expertise of a power system expert.

In order to use (1) and (2) above, one needs to be able to develop a mathematical model of a relevant part of the system. The third method requires that:

(i)   the expertise for solving the problem exists,
(ii) it can be represented with today's technology,

(iii) it can be evoked within a reasonable time span.

These classes of methods have been highly successful in addressing a large set of problems in the power system area. However, there remains a number of problems when the above conditions do not hold. Furthermore, there is a class of problems where the cited approaches lead to unsatisfactory solutions because:

1) the number of computational possibilities is too high leading to large solution times, for example, unit commitment or contingency analysis,

2) they involve tasks that have a statistical character such as load forecasting or signal processing,

3) the model of the relevant part of the system cannot be easily identified, for example, voltage control.

For these classes of systems, Artificial Neural Networks (ANNs), an intelligent machine learning technique, provide a promising and attractive alternative.

There are now very widespread applications of ANNs in a number of different areas of power systems; examples are: load forecasting, security assessment, control, system identification, adaptive autoreclosure, operational planning, protection, etc. In this respect, numerous papers have been written in these areas.

We next consider the application of ANNs in three specific areas: (1) electrical load forecasting, (2) static security assessment and (3) adaptive autoreclosure.

References

1. R P Lippmann, An introduction to computing with neural nets, IEEE ASSP Magazine, 1987, pp.4 - 22
2. S Haykin, Neural networks: a comprehensive foundation, Macmillan College Publishing Company, 1994
3. M A El-Sharkawi, R J Marks II, S Weerasooriya, Neural networks and their application to power engineering, Control and Dynamic Systems, Vol.41, 1991, pp.359 - 460
4. D Niber, Artificial neural networks for power systems: a literature survey, International Journal of Engineering Intelligent Systems for Electrical Engineering and Communications, Vol1, No.3, 1993, pp.133 - 158
5. Y H Song, Y D Han, Neural networks in power systems, Journal of Power System Technology, 1989

CHAPTER 9
NEURAL NETWORKS BASED ELECTRICAL LOAD FORECASTING

9.1 Electrical Load Forecasting

Forecasting electrical load in a power system with lead-times varying from hours to days, has obvious economic advantages. The forecasted information can be used to aid optimal energy interchange between utilities thereby saving valuable fuel costs. Forecasts also significantly influence important operational decisions such as load dispatch, unit commitment and maintenance scheduling. For these reasons, considerable effort is being invested in the development of accurate load forecasting techniques. Load forecasting can be essentially divided into two categories: (i) short term and (ii) long term. Here we will consider short-term load forecasting (24 to 48 hour lead times) only, more so due to the fact that utilities are becoming increasingly dependent upon short-term forecasting for daily energy transactions; this is particularly so because of shrinking energy reserves and the seasonal high energy demand.

Most conventional techniques used for load forecasting can be categorised under two approaches. One treats the load demand as a time series and predicts the load using different time series analysis techniques [1]. The second method is a regression technique which recognises the fact that the load demand is heavily dependent upon weather conditions [2]. However, such traditional techniques often do not give sufficiently accurate results. Conversely, complex algorithmic methods with heavy computational burden can slow the convergence and in certain cases may cause divergence.

A number of algorithms have been suggested for the load forecasting problem. As mentioned before, one approach treats the load pattern as a time series signal and predicts the future load by using various time series analysis techniques. The second approach recognises that the load pattern is heavily dependent upon weather variables and finds a functional relationship between the weather variables and the system load. The future load is then predicted by inserting the predicted weather information into the predetermined functional relationship.

General problems with the time series approach include the inaccuracy of prediction and numerical instability; one reason for the inaccuracy in results is that it does not utilise weather information. In this respect, there is a strong correlation between the behaviour of power consumption and weather variables such as temperature, humidity, wind velocity and cloud cover; this is especially true in residential areas. The time series approach mostly utilises computationally cumbersome matrix-orientated adaptive algorithms which in certain cases may be unstable.

Most regression approaches try to find functional relationships between weather variables and current load demands. In this respect, the conventional regression approaches use linear or piecewise-linear representations for the forecasting functions. By a linear construction of the representations, the regression approach finds the functional relationships between selected weather variables and load demand. However, conventional techniques assume, without any justification, a linear relationship. The functional relationship between changing load demand and weather variables, however, is not stationary but depends upon the non-stationary temporal relationship between the two; the conventional regression approach does not have the versatility to address this temporal variation. It will produce a rather averaged and therefore a very approximate result.

It is apparent from the foregoing that electrical load forecasting is a challenging problem that requires extensive statistical analysis. The problem formulation as well as modelling depends to a large extent on the geographical region where the forecasting is needed. Several key issues must be addressed before a reliable forecast is developed; amongst these are;

    1) The relevant variables with strong correlation to electrical loads such as temperature, clouds, humidity and wind, must be identified.

    2) A reliable feature extraction technique to capture the dominant information related to load patterns and profiles must be developed.

    3) Accuracy of weather forecasting can have a significant impact on the accuracy of load forecasting; in some cases, however, statistical properties of the weather forecasting errors can be captured by the load forecasting model.

    4) One model for load forecasting for all seasons or even all week days may not be possible to be developed with a reasonable degree of accuracy.

5) The forecasting model must be able to extrapolate with a reasonable degree of accuracy during cold snaps, heat waves or pickup loads.

6) The forecasting model must be able to adapt to the system's thermal inertia; it must also be able to handle the load growth.

To include all the aforementioned considerations in the forecasting models is a mammoth task. It may not even be possible to implement without some form of a rule-based system.

9.2 Neural Network Approach

9.2.1 Neural network structure

One very attractive approach to the problem of accurate load forecasting is using a Neural Network (NN) [3,4]; this is so because a NN can combine both time series and regression approaches to predict the load demand. A functional relationship between weather variables and electrical load is not needed. This is so because a NN can technically generate this functional relationship by learning and training data. In other words, the non-linear mapping between the inputs and outputs is implicitly embedded in the NN.

We will now discuss one specific NN-based load forecasting approach which combines both time series and regression approaches [3]. The algorithm utilises an Artificial Neural Network (ANN) based on the Multi-Layer Perceptron (MLP) architecture. As is the case with the time series approach, the ANN traces previous load patterns and predicts (ie extrapolates) a load pattern using recent load data. The algorithm also uses weather information for modelling and the ANN is able to perform non-linear modelling and adaption. One of the principal advantages over conventional techniques is that it does not require the assumption of any functional relationship between load and weather variables in advance. The ANN can be adapted by exposing it to new data.

As discussed in Chapter 8, a ANN can be defined as a highly connected array of elementary processes called neurons. Figure 9.1 is an example of the widely used MLP-type ANN as used in the application considered here. As can be seen, it consists of one input layer, one hidden layer and one output layer. Each layer employs several neurons and each neuron in a layer is connected to the neurons in the adjacent layer with different weights. Signals flow into the input layer, pass through the hidden layer and arrive at the output layer. With the exception of the input layer, each neuron receives signals from the neurons of the previous layer, linearly weighted by the interconnected values between the neurons. The neuron then produces its output signal by passing the summated signal through a sigmoid function of the type discussed in the previous Chapter 8.

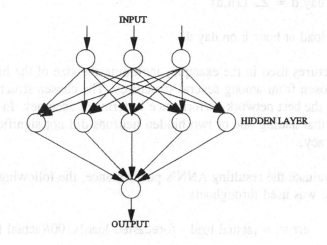

Figure 9.1 Structure of a three-layered perceptron type ANN

A total of Q sets of training data are assumed to be available. Inputs of (i_1, i_2, -----i_Q) are imposed on the top (ie input) layer. The ANN is trained to respond to the corresponding target vector (t_1, t_2, -----t_Q) on the bottom (ie output) layer. The training continues until a certain error-criterion is satisfied. Typically, the training is halted when the average error between the desired and actual outputs of the ANN over the Q training data sets is less than a pre-determined threshold. The training time is dictated by various elements including the complexity of the problem, the volume of data, the network structure and the training parameters used.

9.2.2 Test cases and results

An ANN was trained to recognise the following cases for a typical power supply authority:

- First case: Peak load of the day

- Second case: Total load of the day

- Third case: Hourly load

where

Peak load at day d = max[L(1,d),------L(24,d)]

Total load at day d = $\sum$ L(h,d)

L(h,d) is the load at hour h on day d.

The NN structures used in the example, including the size of the hidden layer, have been chosen from among several structures. The chosen structure was the one that gave the best network performance in terms of accuracy. In most cases, it was found that adding one or two hidden neurons did not significantly affect the NN accuracy.

In order to evaluate the resulting ANN's performance, the following percentage error measure was used throughout:

error = [actual load - forecasted load]x100/(actual load)

First Case

The topology of the ANN for peak load forecasting was as follows:

    Input neurons: T1(k), T2(k) and T3(k)
    Hidden neurons: 5 hidden neurons
    Output neurons: L(k)

where

    k = day of predicted load,
    L(k) = peak load at day k,
    T1(k) = average temperature at day k,
    T2(k) = peak temperature at day k,
    T3(k) = lowest temperature at day k.

Table 9.1 shows the percentage error of each day in the test sets. The average error for all 5 sets is 2.0%.

Second Case

The topology of the ANN for the total load forecasting was as follows:

    Input neurons: T1(k), T2(k) and T3(k)
    Hidden neurons: 5 hidden neurons

Output neurons: L(k)

where

    k = day of predicted load,
    L(k) = total load at day k,
    T1(k) = average temperature at day k,
    T2(k) = peak temperature at day k,
    T3(k) = lowest temperature at day k.

Table 9.2 shows the percentage error of each day in the test sets. The average error for all sets is 1.7%.

Table 9.1 Percentage error of Peak Load Forecasting

days	set1	set2	set3	set4	set5
day1	4.2	1.9	0.7	1.7	1.8
day2	0.2	1.9	3.0	0.3	3.3
day3	0.6	2.4	1.0	2.7	2.7
day4	2.4	3.9	3.3	2.8	1.1
day5	0.4	4.3	0.7	6.6	0.6
day6	2.8	0.1	0.6	1.4	2.0
avg	1.7	2.4	1.6	2.6	1.9

Table 9.2 Percentage error of total load forecasting

days	set1	set2	set3	set4	set5
day1	0.3	0.3	2.7	1.0	0.4
day2	1.0	2.0	1.8	0.7	0.9
day3	3.5	1.0	3.3	0.7	1.4
day4	1.6	1.7	5.6	1.9	2.1
day5	1.0	0.9	4.1	0.0	0.3
day6	1.8	1.1	3.0	1.2	1.1
Avg	1.8	1.1	3.4	1.2	1.0

Third Case

The topology of the ANN for the hourly load forecasting with one hour of lead time was as follows:

    Input neurons: k, L(k-2), L(k-1), T(k-2), T(k-1) and Υ(k)
    Hidden neurons: 10 hidden neurons
    Output neurons: L(k)

where
    k = hour of predicted load,
    L(x) = load at hour x,
    T(x) = temperature at hour x,
    Υ(x) = predicted temperature for hour x.

At the training stage, T(x) was used instead of Υ(X). The lead times of predicted temperatures, Υ(x), varied from 16 to 40 hours.

Table 9.3 shows the percentage error of each day in the test sets. The average error for all 5 sets was found to be 1.4%. Note that each day's result is averaged over a 24-hour period.

Table 9.3 Percentage error of hourly load forecasting with one-hour lead time

days	set1	set2	set3	set4	set5
day1	(*)	1.2	1.4	1.2	(*)
day2	1.7	1.5	(*)	1.6	2.2
day3	1.1	(*)	1.0	(*)	1.7
day4	1.4	1.3	1.4	1.2	1.7
day5	1.3	1.4	(*)	1.2	(*)
day6	(*)	1.5	1.3	1.7	1.0
Avg	1.4	1.4	1.3	1.4	1.6

*: predicted temperatures, Υ, are not available

Figure 9.2 shows examples of the hourly, actual and forecasted load with one-hour and 24-hour lead times as predicted by the ANN, for two separate days in a week.

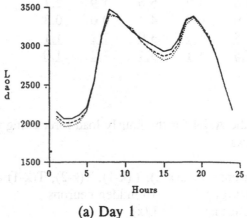

(a) Day 1

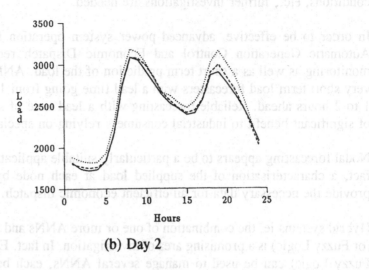

(b) Day 2

Figure 9.2 Hourly load forecasting and actual load in MW (solid: actual load, dash: 1-hour lead forecast, dot: 24-hour lead forecast)

It is apparent from the results that the ANN is well suited to interpolate among the load and temperature pattern data of the training sets to provide the load pattern. In order to forecast the future load from the trained ANN, there is a requirement to use the recent load and temperature data in addition to the predicted future temperature. However, compared to the conventional regression methods, the ANN allows more flexible relationships between temperature and load pattern.

Since the NN simply interpolates among the training data, it will give high error with the test data that is not close enough to any one of the training data. In general, the NN requires training data well spread in the feature space. in order to provide highly accurate results. The training times in the foregoing application considered varied, depending on the cases studied, from about 3 hours to 7 hours CPU time using a high powered workstation. However, a trained ANN required only 3 to 10 ms for testing.

9.3 Conclusions

Numerous feasibility studies have demonstrated the potential of ANNs applied for short term load forecasting. But as load forecasting systems of practical relevance have to cope with different day-type characteristics during the week, varying weather influences, holidays, seasonal changes, abnormal weather conditions, etc., further investigations are needed.

In order to be effective, advanced power system operation functions such as Automatic Generation Control and Economic Dispatch require a real-time monitoring as well as a short term prediction of the load. ANNs can be used as very short term load forecasters with a lead time going from 1 to 15 minutes or 1 to 2 hours ahead. Reliable forecasting with a lead time of several minutes is of significant benefit to industrial consumers relying on special tariffs.

Nodal forecasting appears to be a particularly suitable application for ANNs. In fact, a characterisation of the supplied load at each node by an ANN could provide the necessary data for an efficient economic dispatch.

Hybrid systems ie, the combination of one or more ANNs and an Expert System (or Fuzzy Logic) is a promising area of investigation. In fact, Expert Systems (or Fuzzy Logic) can be used to manage several ANNs, each being devoted to a particular load topology and to adjust the results obtained for rare uncommon days.

References

1. S Vernuri, W Huang, D Nelson, On-line AI-algorithms for forecasting hourly lodas of an electric utitlity, IEEE Trans on Power Apparatus and Systems, Vol.PAS-100, 1987, pp.3775 - 3784
2. C Asbury, Weather load model for electric demand energy forecasting, IEEE Trans on Power Apparatus and Systems, Vol.PAS-97, 1975, pp.1111 - 1116
3. D C Park, M A Sharkawi, R J Mark, Electric load forecasting using an artifical neural network, IEEE Trans on Power Systems, Vol.6, No.2, 1991, pp.441 - 449
4. K Y Lee, Y T Cha, J H Park, Short-term load forecasting using an artifical neural network, IEEE Trans on Power Systems, Vol.7, 1992
5. K Liu, S Subbarayan, R R Shoults, M T Manry, C Kwan, F L Lewis, J Naccarino, Comparison of short-term load forecasting techniques, IEEE Summer Meeting, Paper No 95SM547-0 PWRS, 1995

CHAPTER 10
KOHONEN NETWORKS FOR POWER SYSTEM STATIC SECURITY ASSESSMENT

10.1 Power System Static Security Assessment

Static security assessment is defined as the ability of a power system to reach a state within the specified safety and supply quality following a contingency. Based on estimated data, the static security model evaluates the steady state of the power system neglecting transient and small time-dependent variations.

The main task in power system operation is to decide whether the system is currently operating in a normal state where all constraints on bus voltages, active and reactive bus power and line power flows are respected and whether the system stays secure with respect to an unforeseen outage or fault of m out of n buses, lines, transformers or generators; this means that contingency analysis involves (n-m) contingencies.

There are powerful algorithms such as fast decoupled load flow calculations which treat (n-m) contingencies completely. However, the systematic computation of all failures required by (n-m) security analysis, ie the simultaneous failure of m out of n elements, leads to an exponential increase of the computational time. For large networks, these multiple failures in the electrical network cannot be treated in real time if sequential computers are used to detect insecure situations by simulations. Other methods such as contingency ranking give results but do not detect all insecure states.

Statistical pattern-recognition techniques for security analysis have been proposed to take human experience into account [1]. However, the processing power required for such techniques is very high; the application of associative memories for static security assessment of power system overloads has been proposed. More recently, other artificial intelligence (AI) methods are being studied for the purposes of static security assessment. In this respect, two main approaches to solve problems in power system security analysis using AI techniques can be distinguished. Firstly, expert system techniques that explore human knowledge formulated explicitly as rules [2]; however, such methods are development-intensive and part of the knowledge is system dependent. Secondly, artificial neural network (ANN) based techniques are being increasingly investigated [3,4].

10.2   Neural Network Approach

ANN-based techniques were first proposed for the estimation of critical clearing time for the transient stability problem. In this respect, the applicability of a multi-layer perceptron (MLP) trained with error-back propagation training algorithm to assess static security assessment has been tested on a number of small power systems. However, it has been found that such an approach is time consuming and it requires a vast amount of training data when applied to solving the problem of static security assessment. Moreover, it does not take into account the properties of the system space but tries to conceive the problem as a functional interpolation task; owing to the high dimensionality of the function, accuracy is highly dependent on the number of training points and each contingency has to be viewed as a different function.

The Kohonen's self-organising feature map offers an attractive alternative for the classification of security states of power systems with respect to contingency analysis. As stated in Chapter 8, this type of NN learns similarity of system states in an unsupervised manner; unlike the supervised learning of the error-back-propagation algorithm, unsupervised learning does not require the target class to be specified.

10.2.1 Representation of the operating space

The approach adopted in this example [4] is to view the problem of contingency analysis as a classification task. Here the operating state is chosen to be represented by a vector consisting of the active and reactive power of the transmission lines. This vector is an element of an L-dimensional complex vector space ie, a 2L-dimensional real-valued vector space, called operating space, where L is the number of lines. The operating point of a single contingency is an element of an (L-1) dimensional real-valued subspace.

According to the different security criteria, the boundaries of the secure state space are given by the limits prescribed for the buses and lines. For example, in the case of line overload prevention, these limits are given by the maximum power supported by these lines, ie $|p_{ab} + jq_{ab}| \leq S_{abmax}$, where $p_{ab} + jq_{ab}$ is the complex power flow.

A secure operating point always lies inside the boundaries of the secure space. Critical points lie inside but near the boundaries and insecure points are outside

the boundaries. Figure 10.1 shows the case of the linear power system model
where the secure region of the operating space is convex.

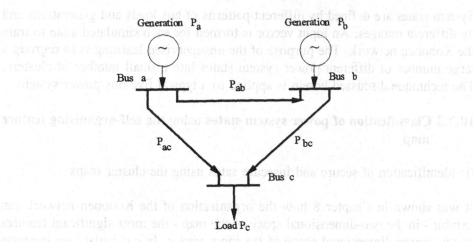

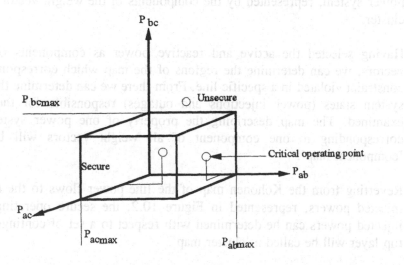

Figure 10.1 Three-bus three-line power system model and the operating space
(the axes represent power between buses a, b, c)

The traditional approach to on-line preventive security analysis is to simulate a
list of outages with respect to the present system state, and to check for each
simulated outage, the branch power flows and voltages against prescribed limits,
in order to assist constraint violations. In contrast, the NN approach is based on

doing the time-consuming simulations of the power system and the training of the NN off-line.

System states are defined by different patterns of bus loads and generations and by different outages. An input vector is formed for each simulated state to train the Kohonen network. The purpose of the unsupervised learning is to regroup a large number of different power system states into a small number of clusters. The technique discussed herein is applied to a typical five-bus power system.

10.2.3 Classification of power system states using the self-organising feature map

(i) Identification of secure and insecure sates using the cluster maps

It was shown in Chapter 8 how the organisation of the Kohonen network can exhibit - in the two-dimensional space of the map - the most significant features of the higher-dimensional space of the input vector. In particular, we interpret the clusters formed by weight vectors in terms of the physical variables of the power system, represented by the components of the weight vectors forming a cluster.

Having selected the active and reactive power as components of the input vectors, we can determine the regions of the map which correspond to a flow constraint violated in a specific line. From there we can determine the prototype system states (power injections and outages) responsible for the constraint examined. The map describing the property of one power system variable corresponding to one component of all weight vectors will be called a 'component map'.

Reverting from the Kohonen map of the line power flows to the maps of the injected powers, represented in Figure 10.2, the secure operating limits for injected powers can be determined with respect to a set of contingencies. This top layer will be called a 'cluster map'.

The high complexity of the security analysis problem is reduced for two reasons. First, the M weight vectors attached to the neurons represent prototypes of the power system states used for training the network. In a real-size power system, this represents a significant reduction of power system states to be analyzed off-line by classical methods. Secondly, the interpolation of the maps for each component of the weight vector indicates which constraints are active for different contingencies.

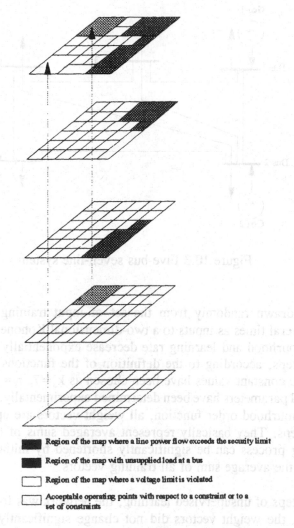

Region of the map where a line power flow exceeds the security limit

Region of the map with unsupplied load at a bus

Region of the map where a voltage limit is violated

Acceptable operating points with respect to a constraint or to a set of constraints

Figure 10.2 Identification of secure and insecure states using the cluster map

This synthesis aspect will now be illustrated in detail for the five-bus seven-line system as shown in Figure 10.3. In this example, the simulation of the Kohonen learning algorithm was set up on a workstation; in this model, all parameters could be defined and changed interactively, such as: the size of the input vector and of the Kohonen network, neighbourhood functions, learning factor α and their decrement with the learning steps.

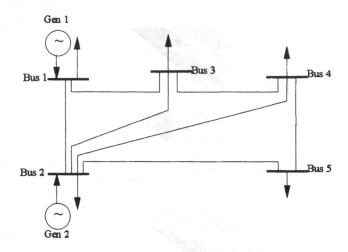

Figure 10.3 Five-bus seven-line system

Vectors are drawn randomly from the set of input training vectors and are presented several times as inputs to a two-dimensional Kohonen network of size 7x7. Neighbourhood and learning rate decrease exponentially with the number of training steps, according to the definition of the functions given in Section 8.2.4(ii). The constant values have been chosen as $k_o = 7$, $\tau_k = 1000$, $\alpha_o = 1$ and $\tau_\alpha = 1500$. All parameters have been determined experimentally. For a large value of the neighbourhood order function, all weight vectors are updated during the first 1000 steps. They basically represent averaged sums of the input vectors. The updating process can be significantly shortened by initializing the weight vectors with the average sum of all training vectors.

After 4000 steps of unsupervised learning, the network was found to be already organised ie, the weight vectors did not change significantly. Weight vectors whose norm was represented by a straight line corresponded to those neurons which were never chosen as the nearest weight vector for any training vector and which therefore did not change any more once the neighbourhood order function was smaller than 1. Further, two weight norms converging to the same stable point might correspond to vectors in different parts of the security region.

(ii) The five-bus seven-line network

Figure 10.3 illustrates the structure of the five-bus seven-line system. The 14 components of the input vector consist of the seven active and reactive line power flows.In the following sections, the classification of the operating points of the 14-dimensional security space by the two-dimensional neural network is studied. Forty-six different line-loading patterns corresponding to the base case and all single and double contingencies have been generated by off-line simulations and have been trained during the learning phase. For the classification phase, the generation and load level of the base case (100%) has been changed uniformly to 90%, 95%, 105% and 110% and the line loading patterns of the corresponding 184 single and double contingencies have been presented to the Kohonen classifier.

(iii) Classification of training data

The classification results for the 46 trained vectors are shown in Figure 10.4. This synthesis representation of the properties of the weight vectors usually called a cluster map. It corresponds to the top layer in Figure 10.2.

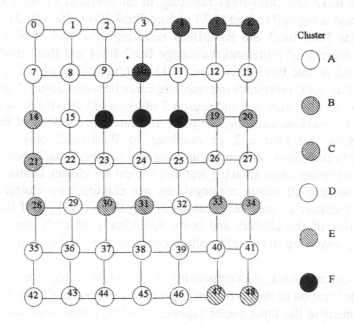

Figure 10.4 Neural cluster map of the classification of 46 trained vectors for the five-bus seven-line system

Each circle of the cluster map represents a neuron. Neurons of the same shade belong to the same cluster. Most clusters are separated by unshaded (white)

neurons. These neurons do not classify any of the input vectors during the classification phase; this means that for every input vector their weight vector has never been chosen as the nearest one.

However, several network states are associated with the same neuron in some cases. For instance, neuron 9 classifies the single outage of line Bus1-Bus2 and the mixed outage of line Bus1-Bus2 and generator Gen1. The line loading patterns of these two cases are very similar.

The clustering of different network states is obvious in some instances. For example, six out of nine cases, single and double contingencies involving the outage of line Bus1-Bus2 are classified by cluster A. These outages provoke an overload of line Bus1-Bus3 and a weak load (less than 30%) of line Bus4-Bus5. The double outage of line Bus1-Bus2 and Bus1-Bus3 results in an islanding of bus1 and is classified by cluster B. The same cluster represents all other cases where the load cannot be totally supplied. Neuron 28 of cluster C classifies outage Bus1-Bus2 and Bus3-Bus4 resulting in an overload of Bus1-Bus3 and Bus2-Bus3 and a negligible load (1%) of line Bus4-Bus5. Neuron 21 classifies outage of line Bus1-Bus2 and Bus2-Bus5 resulting in a load of 60% for line Bus4-Bus5. Neuron 14 represents the outage Bus2-Bus4 and Bus2-Bus5 leading to an overload of line Bus3-Bus4. Cluster D regroups all other outages of line Bus2-Bus3. Cluster C represents intermediate cases between cluster A and cluster D. More specifically, neuron 0 and neuron 7 of cluster D classify those cases not resulting in an overload situation; neuron 1 indicates an overload of Bus1-Bus2 and Bus2-Bus4 and neuron 2 an overload of Bus1-Bus2 only. Cluster F represents all other cases resulting in an overload for Bus1-Bus2. Neuron 1 and 2 therefore represent cases showing features typical for cluster D and cluster F. The base case and all secure contingencies are classified by cluster E. Only neuron 24 of cluster E classifies a case resulting in an overload of Bus2-Bus5. The properties of the clusters and more specifically of each neuron can be identified by analysing the weight vectors component by component.

For a given column index, the components of all weight vectors are presented on a map, at the location of the corresponding neuron defined in Figure 10.4. Since each component of the input vector represents a line power flow and since the weight vectors are weighted sums of the input vectors, each component of the weight vectors characterises the features of the corresponding power flow.

(iv) Generalisation aspects

Figure 10.5 shows the classification results for the 184 untrained vectors. The clusters are now larger. For example, cluster C now connects A and D. Neuron

35, which does not classify any trained vector, now classifies the case of mixed outage Bus1-Bus2 and generator Gen2 outage at load level 90%. The same outages at load level 95%, 100%, 105% and 110% are classified by its neighbour neuron 42. Cluster B still regroups all cases representing islands and unsatisfied loads.

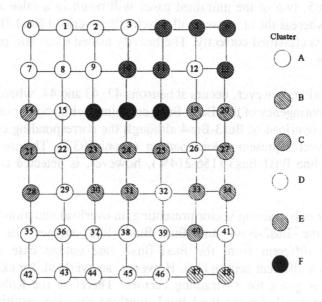

Figure 10.5 Neural network map of the classification of 184 untrained vectors for the five-bus seven-line system

(v) Quality of the classification

The main purpose of the classification process is the determination of overload situations. The Kohonen net not only determines whether a state is secure, critical or insecure but also constructs the weight vector as a prototype for a class of input vectors which then can be analyzed in more detail. As explained in Section 8.2.3(ii), neurons 1 and 2 and all neurons of cluster F classify cases presenting an overload of line Bus1-Bus2. The differences between these cases have been discussed in Section 8.2.3(ii). If the main interest lies only in the detection of overloads and not in the detection of specific features, a less detailed cluster map can be established.

By fixing the limit value to exactly 100% loading, the borders of the insecure cluster F and part of D are given by the 'border neurons' 1, 8, 16, 17, 18, 12 and 13. In the ideal case these neurons will classify overloaded cases and their 'lower' direct neighbours 0, 7, 15, 23, 24, 25, 26 and 27 will classify states heavily loaded ($\leq 100\%$). However, neuron 24, for example, classifies the Bus2-Bus4 and Bus3-Bus4 outage for all five load levels resulting in a load of line Bus1-Bus2 between 83% and 110% and a load for line Bus2-Bus5 between 89% and 112%. Since the weight vector predicts 96% for line Bus1-Bus2 and 100.3% for Bus2-Bus5, two of the untrained cases will result in a false alarm for line Bus2-Bus5 whereas the other two will misclassify overload Bus1-Bus2. Only the trained case is classified correctly. The heavily loaded cases still present critical system states.

Misclassification, however, occurs at neurons 42, 43 and 44, where in four cases the double contingency of line Bus1-Bus2 combined with another outage presents a 107-115% overload of Bus3-Bus4 although the corresponding components of the weight vector present a load pattern of only 93%. The more significant overload of line Bus1-Bus3 (156-214%), however, is detected correctly in all four cases.

Note that the only training vector presenting an overload situation for line Bus3-Bus4 was the Bus2-Bus4 and Bus2-Bus5 line outages. Its features are significantly different from the Bus1-Bus2 line outage case and therefore presented by a different neuron (14). However, an overload, for example for line Bus1-Bus2, is given for 15 training vectors. Therefore the Kohonen network learned more details for the Bus1-Bus2 overload situation resulting in a larger cluster than for the Bus3-Bus4 case.

Table 10.1 summarizes all false alarms and all misclassifications for the 46 trained and 184 untrained cases with respect to overload situations for at least one out of all seven lines. Thus out of 230 cases, 15 are incorrectly classified, but only the eight misclassifications present a danger for the power system. The total error rate is about 6.5% whereas the misclassification rate is only about 3.4%. Note that all neurons are border neurons of an overload cluster. For all misclassifications, they further detect the most significant overload situation correctly. The wrong classification may however occur for a less significant overload of another line.

Table 10.1 Incorrectly classified cases

Neuron	False alarm	Misclassification
1 (insecure)	2	0
18	1	0
14	1	1
21	0	1
24	1	2
28	2	0
42	0	2
43	0	1
44	0	1
Total	**7**	**8**

10.3 Conclusions

Most ANN approaches solve a global task than classical contingency analysis. They give a global description of the operating space (or parts of it) and its security boundaries. However, when working with simulated data, an ANN can only provide an approximation of the supposedly exact non-linear power system model for one specific operating point; the prediction of abnormal conditions is as good as the ones attained through complete contingency analysis. As statistical tools, they are heavily dependent upon a good statistical representation of the operating space.

In the area of static security analysis, supervised and unsupervised ANNs have different objectives. As seen in the aforementioned example, unsupervised approaches are based on dividing the operating space into classes of operating points, thus pre-processing the data set by reducing it into a limited number of typical cases. These cases can then be evaluated using standard methods. Supervised approaches, on the other hand, approximate the security boundaries of the operating space thus memorising data points of a high-dimensional function and interpolating between them. Some prototype systems have been on trial operation in utilities including B.C. Hydro, Canada and the National Grid Company, UK.

References

1. C K Pang, F S Prabhakara, A H El-Abiad, A J Koivo, security evaluation in power systems using pattern recognition, IEEE Trans on Power Apparatus and

Systems, Vol.PAS-93, No.2, 1974, pp.969 -967

2. D J Sobaji, Y H Pao, An airtifical intelligence system for power system contingency creening, Proc IEEE-PICA Conference, 1987

3. M E Aggoune, M A El-Sharkawi, D C Park, M J Danborg, R I Mark, Preliminary results using artifical neural networks for security assessment of power systems, IEEE Trans on Power Systems, Vol.6, No.2, 1991, pp.890 - 896

4. D Neibur, A J Germond, Power system static security assessment using the Kohonen neural network classifier, IEEE-PICA Conference, 1991

5. Y H Song, Y D Han, Q Y Zeng, Framework of fast dynamic security assessment using neural network based pattern recognition, APSCOM, 1991

6. K Demaree, T Athay, K W Cheung, Y Mansour, E Vaahedi, A Y Chang, B R Corns, B W Garrett, An on-line dynamic analysis system implementation, IEEE Trans Power Systems, Vol.9, No.4, 1994, pp.1716 - 1722

7. A R Edwards, K W Chan, R W Dunn, A R Daniels, Dynamic stability screening of electric power systems using artifical neural networks, Proc UPEC'95, 1995

CHAPTER 11
ADAPTIVE AUTORECLOSURE TECHNIQUES

11.1 Autoreclosure Techniques

Any faults occurring on transmission lines, which can for example be caused by lightning, conductor clashing, conductor fouling by vegetation etc., pose a threat to system security and are rapidly removed by the automatic opening of the circuit breakers which terminate the particular circuit concerned. Automatic control of the reclosing of the circuit breakers after initial breaker opening is a long accepted practice which is intended to rapidly restore the transmission system to its normal configuration with the minimum circuit outage time. Experience in operating overhead systems has indicated that typically 80% of line faults are transient in nature and involve insulator flashover and follow-through power arcs across the insulator coordinating gaps. It is however important to note that in practice, even for transient faults, the fault arc path takes a finite time to de-ionize and regain a level of dielectric strength necessary to withstand the application of the full system voltage when reclosure of the circuit breakers re-energizes the line. Unsuccessful reclosure can have dire consequences for the stability and reliability of extra high voltage systems, since a repeated application of faults effectively occurs. Similar considerations apply in respect of sustained (non-transient) faults e.g. those involving fallen conductors, which, though relatively rare, are further re-energized when re-reclosure occurs. The avoidance of reclosure onto permanent faults minimises equipment failure by reducing equipment stresses and torsional vibration/shaft fatigue in generators.

It is also important to note that the time taken for a transient fault arc path to recover full dielectric strength following initial circuit de-ionization, varies very considerably and is by no means constant even for a given line. Some of the more important factors that determine the 'recovery time' are: the strength of capacitive and inductive coupling between the faulted conductor(s) and any adjacent energized HV conductors, the primary arc current duration, the arc length, levels of air pollutant, humidity levels and prevailing with speeds. In most present practice, reclosure is timed to occur at a fixed pre-set time after the initial de-energization which follows a fault, and this so called 'dead time' must be set at the maximum anticipated value to achieve full recovery of the fault path for transient faults. The autoreclosure dead times used in any given application are thus largely a matter of experience. In high speed autoreclosure applications in particular, there are limitations on the extent to which the autoreclosure dead times can be increased to ensure satisfactory reclosure. This is because sustained circuit interruption can induce system instability or can indirectly initiate automatic load shedding caused by low frequency excursions.

In recent years, because of the economic development and regulation, this has created pressure to increase loading on existing transmission networks as an alternative to system reinforcement. This in turn has resulted in a need to reduce transient stability margins together with a concomitant need to reduce autoreclosure dead times to the minimum necessary for successful reclosure. Furthermore, for the above mentioned reasons, there is increasing pressure to avoid autoreclosure altogether for sustained faults.

Transient single-phase-to-ground faults are the most frequent to occur on EHV transmission lines. For such faults, single-pole autoreclosure(SPAR) provides a means of improving transient stability, reducing the torsional impact on generator shafts and reducing the reclosing voltage transients. Furthermore, SPAR [1,2] may become an absolute necessity in applications where construction of additional circuits is not possible. However unsuccessful reclosure with a fixed dead time or reclosure onto a permanent fault may aggravate the potential damage to the system and the equipment as discussed above. In this respect, adaptive SPAR offers many advantages, which include high-speed response to sympathy trips, minimised unsuccessful reclosing, improved stability and a reduction in system and equipment shock in the case of a permanent fault.

Previous field tests of SPAR schemes, laboratory experiments of secondary arcs and computer simulation studies with detailed modelling of the system components all illustrate semi-unique waveform patterns of the voltage transients following initial breaker opening under different system and fault conditions. For example, in principle, the recovery voltage waveforms of the faulted phase can be used to distinguish between permanent and transient faults in the development of adaptive SPAR. However, the actual waveform and magnitude of the faulted voltage are affected by a complex interplay of many factors such as line construction, fault position, prefault loading, source parameters and atmospheric conditions. The nature of the functional relationship between them is extremely complex and therefore this poses many difficulties in the design of adaptive SPAR schemes using conventional approaches. Furthermore, it should be pointed out that in these methods, the permanent fault is assumed to be of the bolted low-impedance type. In practice, the causes of permanent faults vary from system to system and in most cases, they involve some form of high impedance.

As discussed previously, successful application of neural networks (NNs) in other areas of power engineering has demonstrated that they can be employed as an alternative method for solving certain long-standing problems where conventional techniques have had difficulty or have not achieved the desired speed, accuracy or efficiency. This is so by virtue of the adaptive, learning and parallel processing ability of the NN.

In this Chapter, the results of studies of the application of NNs to adaptive SPAR [3] are discussed. Comprehensive studies on the data processing required to set up the training patterns and implementation of the NN itself are described. Firstly, particular characteristics associated with faulted voltage waveforms under various conditions are analyzed. Features of interest are identified and extracted from the faulted waveforms by Fourier Transforms. A three-layer NN is developed and trained by Extended-Delta-Bar-Delta (EDBD) algorithm. The test results of the trained NN presented clearly demonstrate the technical feasibility of NN applications in the design of adaptive SPAR.

11.2   Adaptive Single Pole Autoreclosure

The block diagram of a neural network-based adaptive SPAR scheme is shown in Figure 12.1. The start logic activates the scheme when breaker trip is identified. The signal processing element processes the signal from the capacitor voltage transformer (CVT) and extracts the features from one cycle of voltage waveform for the neural network. The neural network indicates whether the arc extinguishes or not. If not, the features of the next cycle are presented. This procedure is repeated until final arc extinction is identified or until a certain predefined time period has elapsed. The former is then defined as a transient fault, and the latter is defined as a permanent fault. Once the fault is identified to be transient, a signal is sent to reclose the breakers. Otherwise, a signal is sent to trip the other two healthy phase breakers immediately and lock out the SPAR.

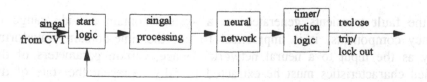

Figure 11.1 Block diagram of neural network based adaptive SPAR

Due to a lack of field data, digital simulation using the Electromagnetic Transient Program (EMTP) is used to generate the sample data required to set up the training/test data for the NN. A typical 400 kV single-circuit transmission line of the type shown in Figure 11.2a with single-pole operation is used for the study. The corresponding system configuration is shown in Figure 11.2b. A very realistic secondary arc model is embedded into the simulation to represent the transient arcing fault. The permanent fault is modelled by linear resistors with realistic values. Various conditions are simulated, which include the effect of: (i) source parameters, (ii) fault location, (iii) fault instant, (iv) prefault loading, and (v) breaker opening time.

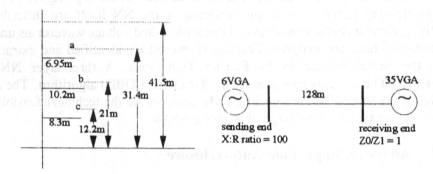

Figure 11.2 The transmission system studied

11.3 Neural Network Approach

The application of neural networks to adaptive autoreclosure schemes consists of four basic tasks: (i) collecting or producing sets of sample of faulted voltage waveforms, (ii) preprocessing the data and extracting the useful features, (iii) choosing and building the most appropriate neural network, and (iv) using the processed sample data to train the neural network and then testing it by simulated fault transient data.

11.3.1 Feature extraction

Since the fault transients generated on a system contain a wide range of frequency components, it is impractical to use the time-domain waveforms directly as the input to a neural network. Hence, certain parameters of the identified characteristics must be extracted to fully represent the state of the transmission line. From an analytical point of view, the most distinct characteristics of the waveforms are those associated with the variation of the frequency components over time. Thus the frequency domain decomposition of features is adopted. The sequential spectral analysis of the faulted voltage waveforms is made by signal processing. A typical example of the spectrum of a voltage waveform for an arcing fault is shown in Figure 11.3b (this is the spectrum of the voltage waveform shown in Figure 11.3a), which clearly illustrates how the frequency components of the waveforms vary with time.

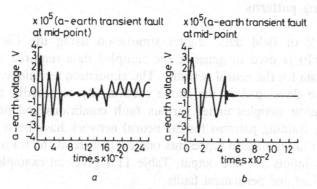

Faulted waveforms of primary system simulation

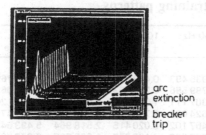

Spectrum of the faulted voltage waveform

x-axis: 1 cm ≡ 2 kHz; y-axis: 1 cm = 20 units; z-axis: 0.5 cm = 225 ms

Figure 11.3 Typical voltage waveform and its frequency spectrum

In order to facilitate the decision making process of the proposed adaptive autoreclosure scheme, the faulted voltage waveforms are separated into sets of one cycle windows. An extensive series of studies using spectrum analysis has shown that for each cycle, certain frequency bands can be used as potential features. In this study, the following six parameters of the faulted voltage have been clearly identified as those representing the most significant features of the state of the transmission lines: (i) fundamental component of faulted voltage; (ii) third harmonic component; (iii) DC component; (iv) second + fourth + sixth harmonic components; (v) fifth + seventh + ninth harmonic components: (vi) components over 500 - 1000Hz range. It should be mentioned that the fundamental frequency, third harmonic and DC components are included individually because of their importance in the overall faulted voltage transients. The other three features which cover the rest of the frequency components of interest, can also be obtained from the bandwidth of the CVT. These six features are then used as the inputs to the neural network. An extensive series of test results have clearly shown the effectiveness of choosing these specific features.

11.3.2 Training patterns

Due to a lack of field data, digital simulation using the Electromagnetic Program(EMTP) is used to generate the sampled data required to set up the training/test data for the neural network. The simulation technique is essentially based on these developed in the above section. By repeatedly analysing the EMTP simulation samples under various fault conditions and processing the resulting data, training patterns for the neural network have been set up. Each vector in the training pattern represents one cycle state of the transmission line. It contains six inputs and one output. Table 11.1 gives an example training set for both transient and permanent faults.

Table 11.1 Example training patterns

DC	50 Hz	150 Hz	100 + 200 + 300 Hz	250 + 350 + 450 Hz	500 Hz + above	Desired
Transient fault						
1.069 924	4.295 334	0.935 493	0.419 045	0.753 492	1.678 765	1
0.659 353	3.996 999	0.799 596	0.210 553	0.748 700	0.376 069	1
0.295 922	4.379 843	1.307 111	1.050 254	1.631 670	1.360 120	1
0.537 762	6.090 184	2.624 032	2.790 713	2.934 785	2.831 471	1
1.986 682	10.814 449	4.467 102	8.020 415	2.518 964	5.493 649	1
4.014 074	18.540 884	1.337 692	4.198 207	1.825 280	2.877 303	1
2.339 997	16.762 474	0.471 079	1.462 525	0.636 406	1.017 475	1
6.847 229	16.155 221	0.093 095	0.277 587	0.104 079	0.160 565	0
5.925 262	16.217 907	0.017 579	0.054 979	0.024 041	0.033 944	0
5.606 986	16.220 527	0.016 308	0.050 998	0.022 268	0.035 624	0
5.310 360	16.222 483	0.015 214	0.047 576	0.020 770	0.033 215	0
5.032 634	16.224 133	0.014 295	0.044 700	0.019 515	0.031 211	0
4.770 965	16.225 564	0.013 502	0.042 222	0.018 433	0.029 480	0
4.523 346	16.226 844	0.012 798	0.040 019	0.017 471	0.027 942	0
Permanent fault						
0.008 609	0.279 653	0.146 066	0.613 925	0.489 286	2.048 051	1
0.054 321	0.150 095	0.001 998	0.006 289	0.002 809	0.007 998	1
0.027 465	0.150 511	0.001 008	0.003 167	0.001 401	0.002 065	1
0.013 359	0.150 602	0.000 539	0.001 685	0.000 734	0.001 177	1
0.005 808	0.150 657	0.000 287	0.000 898	0.000 392	0.000 629	1
0.001 814	0.150 689	0.000 150	0.000 470	0.000 205	0.000 328	1
0.000 250	0.150 707	0.000 076	0.000 238	0.000 104	0.000 166	1
0.001 270	0.150 717	0.000 036	0.000 113	0.000 049	0.000 079	1
0.001 729	0.150 723	0.000 015	0.000 046	0.000 020	0.000 032	1

In window 1 of the transient fault, high values of several frequency components indicate the transition of arc from primary to secondary. The next few windows show the gradual increment of several frequency components, indicating that the arc is 'on' and that the arc length increases with time. In windows 5 and 6, however, marked increases in several frequency components indicate the arrival of the final arc extinction. The subsequent windows show the finally dominant power frequency component associated with the recovery voltage. In the case of a permanent fault, window 1 shows similar high-frequency components. However, the subsequent windows show quite small values of several frequency components due to the discharge through the sustained faulty path to ground. From the foregoing analysis, it can clearly be seen that the desired output of the neural network can be chosen to be '1', to indicate that the fault is still 'on', and '0' to indicate the fault has broken off.

11.3.3 Network architecture and training

The feedforward multi-layer neural network is chosen for this study. From the above analysis, 6 inputs and 1 output were chosen. However, selection of the optimal number of hidden layers, and the optimal number of neurons in each layer are still an open issue. During further studies and analysis, different combinations of the following network training methods were chosen and tested in order to ensure that the model would be continuously refined: (1) different connections (full or non-full) between the processing elements; (2) different number of hidden layers; (3) different hidden neurons in each layer; '(4) different transfer functions (sigmoid, linear, and hyperbolic tangent); (5) different learning set data (sequential or random) in training the network; and (6) different error back-propagation schemes.

Through a series of tests and modifications, it has been found that the network shown in Figure 11.4 provides the best performance. It is a three-layer perceptron, with 6 inputs, 1 output, and 5 neurons in the hidden layer. The transfer function is a Hyperbolic tangent. The network was then trained by randomly processing the training patterns. In this respect, an extensive series of studies have shown that the NN training based on a set of approximately 250 different training patterns is more than adequate to cater for all of practically encountered different system and fault conditions. Furthermore, a comparison between the EDBD and BP learning algorithms have shown that the NN reached the required RMS criterion of 0.01% in approximately 8000 and 25,000 learning iterations for the EDBD and BP algorithms respectively. The former has thus been adopted in this study. As discussed in Chapter 8, the most popular learning algorithm for adjusting the weights for a multi-layer neural network is the back propagation(BP) procedure. It is based on a steepest descent approach to minimize the predication error with respect to the connection weights in the network. Despite its proven effectiveness, the rate of convergence of the procedure is still too slow to be used in a number of practical applications.

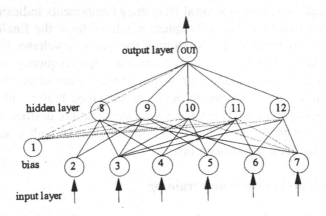

Figure 11.4 Neural network architecture

The Delta-Bar-Delta(DBD) algorithm of Jacobs is an attempt to address the speed of convergence issue, via the heuristic route. Empirical evidence suggests that each dimension of the weight space may be quite different in terms of the overall error surface. Jacobs has proposed a heuristic approach to account for the variation of the error surface, in particular, that every connection of a network should have its own learning rate. This is so because the step size appropriate for one weight dimension may not be appropriate for all weight dimensions. Furthermore, these learning rates should be allowed to vary over time. In enhancements to Jacobs' DBD learning rule, Minai and Williams have proposed an extended-DBD algorithm(EDBD) which incorporates momentum adjustment, based on heuristics, to increase the rate of learning and to damp the oscillations. It is the EDBD algorithm which is adopted in this study.

Consider a feedforward network, in which the basic BP algorithm adjusts the weights to minimize the total squared error, E. The latter is the sum of the difference squared between the set of the desired outputs d_k and the set of actual outputs o_k of the NN and is given as:

$$E = \sum (d_k - o_k)^2 \qquad (11.1)$$

The connection weight update given by the standard Delta-Rule with error function E(k) is:

$$\Delta w(k+1) = \alpha \delta(k) + \mu \Delta w(k) \qquad (11.2)$$

also,

$$w(k+1) = w(k) + \Delta w(k+1) \qquad (11.3)$$

where learning rate α and momentum rate μ are constants. For EDBD, the time-varying learning rate $\alpha(k)$ and time-varying momentum rate $\mu(k)$ are assigned to each connection in the network as follows:

$$\Delta w(k+1) = \alpha(k) \delta(k) + \mu(k) \Delta w(k) \qquad (11.4)$$

and, as before

$$w(k+1) = w(k) + \Delta w(k+1) \qquad (11.5)$$

The momentum rate $\mu(k)$ and the learning rate $\alpha(k)$ are adjusted according to the rules below.

$$\delta(k) = (1-\theta)(\delta(k) + \theta \delta(k-1)) \qquad (11.6)$$

The learning rate change for EDBD is:

$$\Delta\alpha(k) = \begin{cases} K_\alpha \exp(-\gamma_\alpha |\delta(k)|) & \text{if } \overline{\delta}(k-1)\delta(k) > 0 \\ \varphi_\alpha \alpha(k) & \text{if } \overline{\delta}(k-1)\delta(k) < 0 \\ 0 & \text{otherwise} \end{cases} \qquad (11.7)$$

and the momentum rate change is, similarly defined as:

$$\Delta\mu(k) = \begin{cases} K_\mu \exp(-\gamma_\mu |\delta(k)|) & \text{if } \overline{\delta}(k-1)\delta(k) > 0 \\ \varphi_\mu \mu(k) & \text{if } \overline{\delta}(k-1)\delta(k) < 0 \\ 0 & \text{otherwise} \end{cases} \qquad (11.8)$$

The learning and the momentum rates have separate constants controlling their increase and decrease. The polarity of $\overline{\delta}(k)\delta(k)$ is used to indicate whether, heuristically, an increase or decrease is appropriate. To prevent wild jumps and oscillations in the weight space, ceilings are placed on the learning rates and momentum rates, i.e.:

$$\alpha\,(k) \leq \alpha_{max} \qquad \mu\,(k) \leq \mu_{max} \qquad\qquad (11.9)$$

It should also be noted that in order to prevent wild jumps and erratic oscillations in the weight space, ceilings of 1.8 and 0.9 have been placed on the learning and momentum rates respectively. It has been found that these values always ensure that the NN gives the desired performance.

11.4 Test Results

11.4.1 Typical NN performance

The trained NN was tested using a set of test data generated from the fault simulator, in the same way as the data used for training the NN. Table 11.2 illustrates an example of the test results attained for both transient and permanent faults, both being a-earth faults at the midpoint of the line for the system configuration shown in Figure 12.2. A comparison between the desired output and the NN outputs clearly shows that the results attained from the latter are very satisfactory, in that they are very close to the ideal outputs of either 'zero' or 'one'. As mentioned previously, a change from near 'zero' to near 'one' indicates a transient fault and a no change constitutes a permanent fault. Furthermore, in the case of a transient fault, the time at which the change occurs, gives the precise arc extinction time. In practice, since there is always a small fluctuation of the NN output around 'one' and 'zero' (as shown in Table 11.2), small threshold levels have to be set. For example, if the output falls within the range ($< |0.1|$) then this would be classified as a 'zero' i.e. fault-off and if it falls within a range (0.8 - 1), it can be classified as a 'one' i.e. fault-on. Figure 11.5 gives a graphics representation of the outputs of NN for faults at sending and remote ends, respectively.

When a high impedance fault is presented to the NN which is not trained for this type of fault, the results, as shown in Table 11.3, are very good. However, in order to minimize the probability of any adverse decision from the NN, it is vitally important to train the network with as much data as possible attained from simulation studies, field tests, etc.

Table 11.2 Example test results

Transient fault		Permanent fault	
Desired	Output of NN	Desired	Output of NN
Secondary arc		1.0	0.999 067
1.0	0.999 046	1.0	0.950 273
1.0	0.993 631	1.0	0.950 227
1.0	0.999 105	1.0	0.950 325
1.0	0.999 105	1.0	0.979 530
1.0	0.999 294	1.0	0.995 999
1.0	0.999 298	1.0	0.871 082
1.0	0.932 143	1.0	0.873 124
1.0	0.960 058	1.0	0.874 754
		1.0	0.875 601
Arc extinction		1.0	0.876 039
0.0	−0.006 467	1.0	0.876 277
0.0	−0.008 551	1.0	0.875 947
0.0	−0.006 966	1.0	0.875 837
0.0	−0.004 656		
0.0	−0.001 388		
0.0	0.003 126		
0.0	0.009 240		

(Permanent fault path)

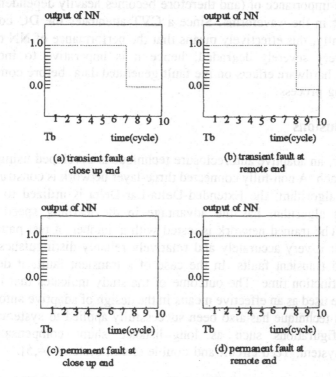

(a) transient fault at close up end

(b) transient fault at remote end

(c) permanent fault at close up end

(d) permanent fault at remote end

Figure 11.5 Graphical output of the NN in a number of test cases

Table 11.3 Test results of a high impedance fault

	Permanent high impedance fault path				
Desired	1.0	1.0	1.0	1.0	1.0
Output of NN	0.963 394	0.948 631	0.948 612	0.998 889	0.977 867

11.4.2 Effect of noise on NN performance

All the foregoing results presented are for cases when the NN is subjected to faulted voltage waveforms which include elements of the previously mentioned noise. With regard to the effect of the white noise component, results have shown that the latter has no adverse effect on performance and this can be directly attributed to robustness. With regard to the effect of CVT errors, a series of tests performed earlier in the work showed that if the training process was based on raw primary system waveforms and the NN was then subsequently subjected to voltage outputs from the CVT, it gave erratic decisions. This is so because in the case of the training based on primary system data, the NN accentuates the importance of (and therefore becomes heavily dependent on) the DC component in the waveforms. Since a CVT attenuates the DC component quite significantly, this effectively means that the performance of NN trained in this way is very severely degraded; hence it is imperative to include the transducer and hardware effects on the fault generated data, before commencing the NN training process.

11.5   Conclusions

In this Chapter, an adaptive autoreclosure technique is developed using a neural network approach. A non-fully connected three-layer network is constructed, and as a learning algorithm, the Extended-Delta-Bar-Delta is utilized to train the network. This algorithm has the advantage in the learning speed and fast convergence. The trained network is tested with a number of test patterns, and it is found that it very accurately and relatively reliably distinguishes between permanent and transient faults. In the case of a transient fault, it defines the precise arc extinction time. The outcome of the study indicates that the neural network can be used as an effective means in the design of adaptive autoreclosure schemes. This technique has also been successfully applied to systems of more complex configurations such as long-distance shunt compensated EHV transmission system, Teed circuit and double circuit systems [4,5].

Because field data describing the phenomena of the transient state of the transmission line is normally very difficult to obtain, realistic digital simulation becomes essential. Simulation techniques previously developed are therefore successfully used to generate the sample data for this study. The analyses of the associated characteristics of the faulted voltage waveforms provide a basis for extracting the appropriate features to transform state samples into concise and meaningful parameters. This, in turn, simplifies the architecture of the neural network.

The results clearly show that spurious noise generated within the measured signals as a result of the dynamic nature of the power system has no adverse effect on NN performance. However the effect of transducer (CVTs) and hardware errors need to be taken into account when training the NN, since they can have a significant bearing on the learning methodology adopted during the training procedure.

Furthermore, the proposed technique has been tested against the field recorded data. The prototype hardware has been on trial operation in ESCOM of South African.

References

1. R K Aggarwal, Y H Song, A T Johns, Adaptive single-pole autoreclosure scheme based on defining and identifying fault induced voltage waveform patterns, Proc ATP, Greece, 1993
2. Y Ge, F Sui, Y Xiao, Prediction methods for preventing single-phase reclosing on permanent faults, IEEE Trans on Power Delivery, Vol.4, No.1, 1989
3. R K Aggarwal, A T Johns, Y H Song, R W Dunn, D S Fitton, Neural-network based adaptive single-pole autoreclosure technique for EHV transmission systems, Proc IEE GTD, Vol.141, No.2, 1994, pp.155 - 160
4. Y H Song, R K Aggarwal, R W Dunn, D Fitton, neural network based adaptive autoreclsoure for teed circuits, Proc IEEE/CSEE PSTC, 1994
5. R K Aggawal, Y H Song, A T Johns, Adaptive three-phase autoreclosure for double-circuit systems using neural networks, Proc APSCOM, 1993
6. D S Fitton, R W Dunn, R K Aggarwal, A T Johns, A Bennett, Design and implementation of an adaptive single pole autoreclosure technique for transmission lines using artificial neural networks, IEEE Summer Meeting, 1995

CHAPTER 12
GENETIC ALGORITHMS: AN INTRODUCTION

12.1 Characteristics of Genetic Algorithms

Based on the principles of genetics and natural selection - Darwin's "survival of the fittest" strategy -Genetic Algorithms (GA)s are adaptive search techniques. The underlying principles of Genetic Algorithm were first reported by Holland in 1962. Genetic Algorithm simulates a heuristic probabilistic search technique that is analogous to the biological evolutionary process. The process possesses a feature that species with the higher fitness function to environments are able to survive and evolve over many generations while those with the low fitness function are unable to survive and eventually become extinct due to the natural selection. Chromosomes play an important role in the process. The probability that species survive and evolve to prosper over generations becomes higher as the fitness increases. In the words of optimization, the solution approaches the optimum.

The features of genetic algorithms are different from other search techniques in several aspects:

(i) GAs optimize the trade-off between exploring new points in the search space and exploiting the information discovered thus far.
(ii) GAs have the property of implicit parallelism. Implicit parallelism means that the GA's effect is equivalent to an extensive search of hyperplanes of the given space, without directly testing all hyperlane values.
(iii) GAs are randomized algorithms, in that they use operators whose results are governed by probability. The results for such operations are based on the value of a random number.
(iv) GAs operate on several solutions simultaneously, gathering information from current search points to direct subsequent search. Their ability to maintain multiple solutions concurrently makes GAs less susceptible to the problems of local maxima and noise. A comparison between single and multiple path search is illustrated in Figure 12.1.

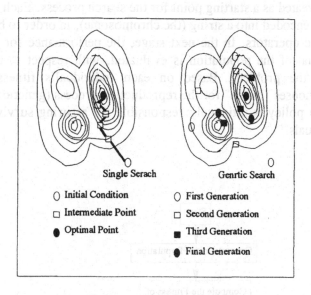

Figure 12.1 Single search and multiple search

12.2 A Simple Genetic Algorithm

A Genetic Algorithm is a computational model that emulates biological evolutionary theories to solve optimisation problems. A GA comprises a set of individual elements (the population) and a set of biologically-inspired operators defined over the population itself. According to evolutionary theories, only the most suited elements in a population are likely to survive and generate offspring, thus transmitting their biological heredity to new generations.

In computing terms, a genetic algorithm maps a problem on to a set of (binary) strings, each string representing a potential solution. The GA then manipulates the most promising strings searching for improved solutions. A GA operates typically through a simple cycle of four stages:

i) creation of a set of string,
ii) evaluation of each string,
iii) selection of the "best" strings, and
iv) manipulation to create the new set of strings.

Figure 12.2 shows these four stages using the biologically inspired GA terminology. In each cycle, a new generation of possible solutions for a given problem is produced. At the first stage, an initial population of the potential solutions is created as a starting point for the search process. Each element of the population is encoded into a string (the chromosome), in order to be manipulated by the genetic operators. In the next stage, the performance (or the fitness) of each individual of the population is evaluated, with respect to the constraints imposed by the problem. Based on each individual's fitness, a selection mechanism chooses "mates" for the reproduction process (genetic modification). The selection policy is ultimately responsible for assuring survival of the best fitted individuals.

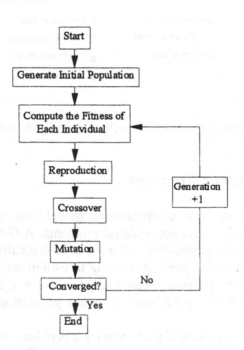

Figure 12.2 The concept of genetic algorithm

12.2.1 Coding and generation of initial populations

GAs work with a coding of the parameter set, not the parameters themselves. The basic coding technique is to use a binary representation. In addition, there are several coding methods with alphabets and figures. What is important in coding is to use more simplified coding.

Then the initial populations with appropriate size are generated by random numbers.

12.2.2 Selection and reproduction

The reproduction process employs genetic operators to produce a new population of individuals (offspring) by manipulating the "genetic information" according to their fitness function values, referred to as genes, possessed by members (parents) of the current population. Eliminating strings according to their fitness function values means that strings with a lower value have a lower probability of contributing and conversely, coping strings according to their fitness values means that strings with a higher value have a higher probability of contributing.

12.2.3 Crossover

The crossover operation is used to create a new individual with a pair of parents. Crossover takes two chromosomes and swaps part of its genetic information to produce new chromosomes. This operation is analogous to sexual reproduction in nature. It is important to create children without killing better parents. There are several crossover schemes such as (1) one point crossover; (2) two point crossover and (3) uniform crossover. For example, Figure 12.3 illustrates a typical one point crossover operation. After the crossover point has been randomly chosen, the portions of strings Parent 1 and Parent 2 are swapped to produce the new strings Child 1 and Child 2.

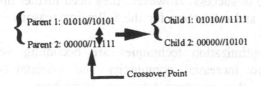

Figure 12.3 Example of one-point crossover

12.2.4 Mutation

Mutation is the other important genetic operator. Mutation may introduce new genetic information into the population. It is implemented by occasionally altering a random bit in a string. With the binary string representation, this simply means changing a "1" to a "0" and vice versa. For instance, Figure 12.4 shows the mutation operator being applied to the third element of the string.

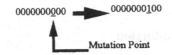

Figure 12.4 Example of mutation

The selection and crossover can normally cover the entire space of operation. In some instances, some potentially useful genetic material may be lost, and mutation maybe needed. Mutation is by itself a random walk through the string space to provide for occasional disturbances in the crossover operation. When used sparingly with selection and crossover, it ensures diversity in the genetic strings over long periods of time, and prevents stagnation in the evolution of optimal individuals.

The creation-evaluation-selection-manipulation cycle is repeated until a satisfactory solution to the problem is found.

12.3 Genetic Algorithms in Power Systems

Optimization is a very important and complex component of power system planning and generation. It involves a large variety of considerations. Conventional optimization techniques, such as nonlinear, linear, quadratic, and dynamic programming , have all been applied to power optimization problems with a certain degree of success. However, they need further improvement with respect to efficiency and robustness for the following reasons:

(1) Conventional optimization techniques are becoming very complicated formulations due to increased complexity and uncertainty arising from environmental issues, thus exposing deficiency in problem solving techniques.
(2) Conventional optimization techniques have to sometimes decompose a highly integrated problem into several more manageable subproblems and solve them sequentially, which greatly increases the complexity of the problem and reduces the efficiency.
(3) Conventional optimization techniques are susceptible to unforeseeable constraints and contingencies in the field.

Genetic algorithms have emerged as an attractive alternative or complement to Conventional optimization techniques. GAs with an inherent global optimization property, offer a fast, robust and efficient algorithm. Since it was first introduced to solve reactive power scheduling in 1991, many papers have appeared which study the feasibility and capability of GAs over a broad range of power system problems. The main areas include: (1) economic dispatch; (2) unit commitment; (3) reactive optimization; (4) planning and (5) power system control. In the following Chapter, an example in environmental-constrained economic dispatch will be introduced.

References

1. D E Goldberg, Genetic algorithms in search, optimization and machine learning, Addison-Wesley Publishing Company, 1989
2. F Li, Y H Song, R Morgan, D T Y Cheng, Genetic algorithms in power system optimization, Proc Conf on Adaptive Computing, 1994
3. J H Hooland, Adaptation in natural and artificial systems, The University of Michigan Press, 1975
4. The Proceedings of the International Conference on Genetic Algorithms and Their Applications, 1985 (the first), 1987(the second), 1989 (the third), 1991 (the fourth)

CHAPTER 13
GENETIC ALGORITHM IN POWER SYSTEM OPTIMIZATION

13.1 Environmentally Constrained Power Economic Dispatch

Under the ever strict governmental regulations on environmental protection, the conventional operation at absolute minimum cost can no longer be the only basis for dispatching electric power. Society demands adequate and secure electricity not only at the cheapest possible price, but also at minimum levels of pollution. This requires a new dispatch problem structure to be formulated in a power plant to give simultaneous treatment of minimisation of cost and abatement of emission. The goal of new power dispatch thus becomes to schedule the generator output to supply the customer load demand while minimizing environmental impact at the minimum operating cost.

A reduction of the SO_2 emission can be achieved either by system plant-level redesign or by changing operational strategy. The simplest way is to change system dispatch strategy by including environmental considerations, either as a second objective or as an additional constraint into conventional economic dispatch. The result is known as Environmental-Economic Dispatch (EED) which is an attractive tool for reducing both emission and cost, as it requires least lead time and capital investment. However, it has the disadvantage that its operational capacities are limited. As a result, the ability of EED to reduce emission is restricted. The utilities have to make great efforts to change power generation equipment to meet ever increasing environmental regulations. To reduce such time and cost consuming plant level changes, a fuel switching (FS) technique can be used to obtain a compromise between system hardware redesign and software redesign.

This chapter employs both EED and FS techniques to optimize fuel mixture and minimize fuel cost under environmental constraint in order to maximize the emission reduction capacity without plant level action. To help find the best solution to the problem of achieving both optimal fuel mix and minimum possible fuel cost, the chapter proposes a two-phase dispatch structure to approach the problem. In the first phase, the problem is modified to a multi-optimization function, where unit power output and high sulphur and low sulphur fuel ratio are two vector variables. With the results obtained from the first phase, the second phase, which contains only a single cost criterion, searches for the optimal solution in detail in the optimal sulphur region. The essence of the proposed method is that phase one provides the optimal search region for phase two to work with, essentially to obtain a better overall solution.

Conventional techniques have drawbacks in approaching the above bi-optimization problem. These techniques start their search at a single point, and gradually move it towards the optimal point. In contrast, a genetic based algorithm (GA), as employed in this chapter, is forced to search for the optimum point from a group of points in order to reach a set of feasible solution from many aspects of the problem. The mechanism of a GA resembles the process found in natural evolution - survival of the fittest.

The attraction of a GA lies in its computational simplicity and its problem solving efficiency, particularly in cases where conventional techniques have not achieved the desired speed, accuracy or efficiency. Various results have demonstrated that GAs offer robust and efficient strategies for a broad range of power system optimization. In this chapter, a Genetic Algorithm has been used to attack both multiple and single optimization problems imposed by the two-phase EED structure. The solution method can attack the problem efficiently, and it is guaranteed to find the optimal point or a solution which is near to it.

13.2 Genetic Algorithm Approach

13.2.1 EED Strategy with Fuel Switching

The classic power dispatch problem, where the environmental consideration is absent, aims to supply the required quantity of power at the lowest possible cost. The dispatch problem can be stated mathematically as follows:

To minimise the total fuel cost at thermal plants:

$$\text{minimize } F = \sum F_i(P_i) \tag{13.1}$$

Subject to the equality real power balance constraint:

$$\varphi = \sum P_i - P_L - P_D = 0 \tag{13.2}$$

and the inequality constraint of limits on the generator outputs is:

$$Pmin < P < Pmax \tag{13.3}$$

where n is the number of generators committed to the operating system, $F_i(P_i)$ is the individual generation production cost in terms of its real generation power P_i. P_L represents the transmission losses and P_D is the constant load demand.

The proposed EED strategy takes both economic and environmental priorities into account, as well as the employment of fuel switching, the overall problem formulated is stated as:

$$\text{minimize } F_s = \sum F_i(P_i, S_i) \tag{13.4}$$

subject to the same equality and inequality constraints (equations (13.2-13.3) as the classic ED problem. The fuel cost function can be represented by multiplying the quantity of fuels consumed by the cost, which is:

$$F_i = (a_i - b_i S_i)(\alpha_i + \beta_i G_i + \gamma_i G_i^2) \tag{13.5}$$

where a_i, b_i are the cost per kcal coefficients, α_i, $бi$, γ_i are fuel consumption coefficients.

The variable S_i is the sulphur content in the fuel for unit i. The sulphur in the fuel for each unit is decided by the mix ratio of high and low sulphur fuels. Assuming the percentage of high sulphur content $SU_i(\%)$ is $W_i(W_i < 1.0)$, the low sulphur fuel with $SL_i(\%)$ then becomes $(1-W_i)$. The overall sulphur in the fuels can thus be expressed as:

$$S_i = SU_i \times W_i + SL_i \times (1 - W_i) \tag{13.6}$$

The additional inequality constraint introduced by fuel switching is the upper and lower bounds upon sulphur in fuels:

$$SL_i \leq S_i \leq SU_i \tag{13.7}$$

SL_i, SU_i are the maximum and minimum sulphur contents in fuels respectively.

The emission limit for the area is as:

$$Q = Q_{AREA} - \sum Q_i > 0 \tag{13.8}$$

where the emission for each unit is:

$$Q_i = e_i S_i (\alpha_i + \beta_i G_i + \gamma_i G_i^2) \tag{13.9}$$

where Q_i is the SO_2 emission per hour for unit i, Q_{AREA} is the total permissible emission per hour.

13.2.2 Design of GA for Two-phase Problem Structure

The fitness function is the most important factor in GA implementation, as it is the only information to guide the search towards the optimum point. This section employs a novel fitness function formulation. The fitness is formed by normalisation of the objective functions and the constraints, so that it can not only treat objective and constraint equally, but can consider additional constraints very easily. The fitness function for the proposed two-phase EED problem is described as:

$$FN = Fs/(Fsmax-Fsmin) + \lambda\Phi/(\Phi max-\Phi min) + \mu\theta/(\theta max-\theta min) \tag{13.10}$$

Where Fs, Φ and θ are the fuel cost objective, the equality power balance constraint and the area emission constraint respectively. The corresponding weight coefficients are λ and μ.

As for the strings to encode the problem, the string length is chosen as 10 bits per unit for both phases. In the phase one, the first 5 bits represent power output, and the latter 5 bits represent sulphur content in fuel, while in phase 2, as only one variable is present, all 10 bits are used for power output, which means that the space searched is doubled. This gives the possibility of obtaining a better result.

Other GA parameters are chosen as: total generation is 30; population size as 100; crossover probability as 0.9 and mutation rate as 0.01.

For both phases, the inequality constraints for control variables of equations (13.2), (13.7) have been handled in the decoding section. Binary coding is used for both phases to effectively explore the search space.

13.2.3 Simulation Results

The proposed dispatch policy tries to decide the load dispatch and the sulphur contents for each unit in order to get desired cost reduction under the environmental constraints. Digital simulation studies by genetic algorithm have been carried out on a system with four generator units [2]. For comparison purposes, the conventional one phased structure, which contains only the first phase of the proposed structure has also been tested on the example system. The daily load curve is shown in Figure 13.1, and the permissible limit on the total emission per hour is $800Nm^3/h$. Figure 13.2 is the cost comparison result for one phase and two phase problem structure, where phase two shows the obvious cost improvement over the conventional one phased structure. Table 13.1 gives the detailed figure of cost improvement achieved using the two phase problem structure proposed.

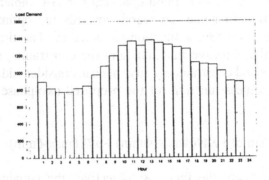

Figure 13.1 Daily load demand curve

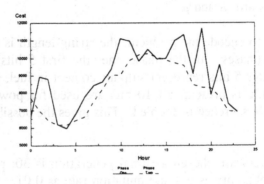

Figure 13.2 Two-phase cost curve with GAs

Table 13.1 Comparison of the total fuel cost

	Total Fuel Cost (24 hours)	Cost Ratio
One phase	19,8749	100
Two phase	19,7435	99.5

Figure 13.3 is the calculated power outputs for each unit from the two-phase structure by Genetic Algorithms. Figure 13.4 is the sulphur content for each unit from the phase one by the same genetic based technique.

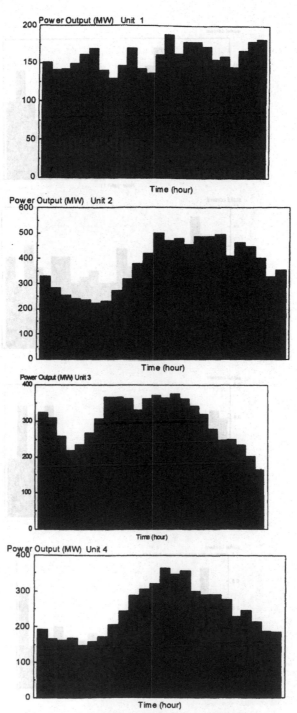

Figure 13.3 Resulting power outputs

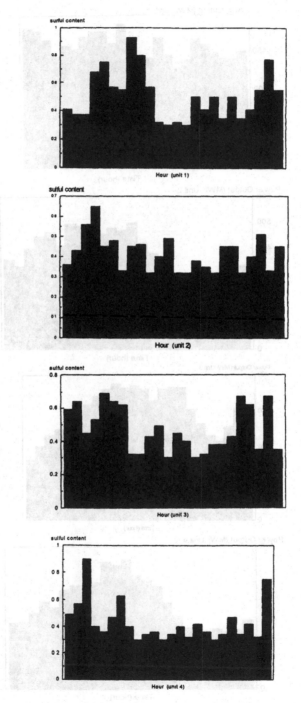

Figure 13.4 Resulting surfer contents

13.3 Conclusion

This chapter proposes a new two-phase problem structure for economic-environmental power dispatch, where both economic-emission dispatch and fuel switching techniques are employed. The proposed method is tested against a conventional one phase problem structure, and is proven to be more efficient. In order to obtain better results, the paper improves the performance further by employing Genetic Algorithms. GAs are proved to be capable and suitable in both multi-objective and single-objective problems, and thus better than or comparable to conventional optimization techniques. The simulation results successfully demonstrate the feasibility and potential application of GAs in power optimization.

References

1. Y H Song, F Li, R Morgaa, D Wiiliams, Environmentally constrained electric power dispatch with genetic algorithms, Proc IEEE ANN/EC, 1995
2. A Tsuji, Optimal fuel mix and load dispatching under environmental constraints, IEEE Trans Power App. and Syst., Vol.PAS-100, No.5, 1981, pp.2357 - 2364
3. Y H Song, F Li, R Morgan, D T Y Cheng, Comparison studies of genetic algorithms in power system economic dispatch, Power System Technology, Vol.19, No.3, 1995, pp.28 -38
4. K Britting, G B Sheble, Refined genetic algorithm - economic dispatch example, IEE Winter Meeting, 1994

CHAPTER 14
INTEGRATION OF FUZZY LOGIC, NEURAL NETWORKS AND GENETIC ALGORITHMS

14.1 Motivation for Hybrid Systems

Many intelligent system techniques have been developed over the last decade. Some of the major ones include expert systems, fuzzy systems, neural networks and genetic algorithms, which have been introduced in the previous chapters. The applications of these intelligent techniques to power systems have been demonstrated through those chapters. Because of the nature of various types of power system problems, different types of solution may be required. The real world power system problems may neither fit the assumptions of a single technique nor be effectively solved by the strengths and capabilities of a single technique. One approach to deal with these complex real world problems is to integrate the use of two or more techniques in order to combine their different strengths and overcome each other's weaknesses to generate hybrid solutions.

In recognition of the drawbacks associated with purely symbolic, numeric or distributed AI-based representations for dealing with complex and real-world problems, it is now becoming apparent that the integration of various intelligent techniques as shown in Figure 14.1 is a very important way forward in the next generation of intelligent system. The hybrid system techniques can be developed in a variety of ways. Two general approaches can be used to design hybrid systems: function-replacing integration (where a technology is embedded into another) and intercommunicating integration (whereby the respective technologies maintain their respective functional forms, but share data through different degrees of coupling). In the following sections, some of those two types of hybrid systems will be introduced. Example applications in power system will be described.

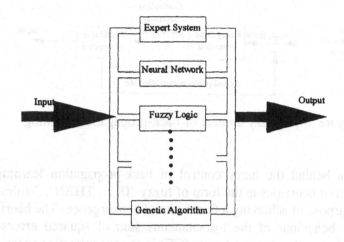

Figure 14.1 Hybrid intelligent systems

14.2 Some Hybrid Intelligent Techniques

14.2.1 Fuzzy-neural networks

There are two approaches to fuzzy neural architecture which we refer to as neural-centric and fuzzy-centric. Neural networks are used as ancillary tools either to determine membership functions, or to change the set of rules adaptively. The neuro-centric approach gives primacy to the neural side of the description. Fuzzy arithmetic is used as a way of improving neuronal and/or entire network behaviour. The fuzzy-centric approach emphasizes a fuzzy logic description, i.e., the main algorithm is formulated through fuzzy IF-THEN rules.

(i) Fuzzy control learning algorithms for back-propagation

Although back-propagation is one of the most popular neural network algorithms, deficiencies such as the convergence speed and local minimum have been identified. In this resect, a number of techniques have been proposed to improve the standard BP. Among them, fuzzy logic controller has become a promising alternative. Figure 14.2 presents the block diagram of a hybrid learning system, in which an on-line fuzzy logic controller is used to adapt the learning parameters of a multilayer perceptron with back-propagation learning. The objective here is to provide a significant improvement in the rate of convergence of the learning process.

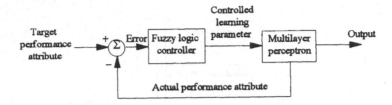

Figure 14.2 Fuzzy control of back-propagation learning

The main idea behind the fuzzy control of back-propagation learning is the implementation of heuristics in the form of fuzzy "IF..., THEN..." rules that are used for the purpose of achieving a faster rate of convergence. The heuristics are driven by the behaviour of the instantaneous sum of squared errors. In the following heuristics, the change of error (CE) is an approximation to the gradient of the error surface, and the change of CS (CCE) is an approximation to the second-order gradient information related to the acceleration of convergence:

(1) IF CE is small with no sign changes in several consecutive iterations of the algorithm, THEN the value of the learning-rate parameter should be increased.
(2) IF sign changes occur in CE in several consecutive iterations of the algorithm, THEN the value of the learning-rate parameter should be decreased, regardless of the value of CCE.
(3) IF CE is small AND CCE is small, with no sign changes for several consecutive iterations of the algorithm, THEN the values of both the learning-rate parameter and the momentum constant should be increased.

A fuzzy rule base for fuzzy BP learning is given in Tables 14.1 and 14.2. The associated membership functions, labelled by linguistic terms, are shown in parts (a) and (b) of figure 14.3. Table 14.1 and Figure 14.3(a) pertain to change applied to the learning-rate parameter η of the back-propagation algorithm, whereas Table 14.2 pertains to changes applied to the momentum constant α of the algorithm. The details of these two tables are as follows:

(1) The contents of Table 14.1 represent the value of the fuzzy variable $\Delta\eta$, denoting the change applied to the learning-rate parameter η, for fuzzified values of CE and CCE. For instance, we may read the following fuzzy rule from Table 14.1: IF CE is zero, AND IF CCE is negative small, THEN $\Delta\eta$ is positive small.

(2) The contents of Table 14.2 represent the value of the fuzzy variable $\Delta\alpha$, denoting the change applied to the momentum-rate parameter α, for fuzzified values of CE and CCE. For instance, we may read the following fuzzy rule from Table 14.1: IF CE is negative small, AND IF CCE is negative small, THEN $\Delta\alpha$ is zero.

From the membership function shown in Figure 14.3 we note that the universe of discourse for both CE and CCE is [-0.3,0.3]; values of CE and CCE outside of this range are clamped to -0.3 and 0.3, respectively.

Having determined whether $\Delta\eta$, the change in the learning-rate parameter, is positive small, zero, or negative small, we may then assign an appropriate value to the change $\Delta\eta$ using the membership functions presented in Figure 14.3.

Table 14.1 Decision table for the fuzzy control of learning rate parameter

CCE	CE				
	NB	NS	ZE	PS	PB
NB	NS	NS`	NS	NS	NS
NS	NS	ZE	PS	NE	NS
ZE	ZE	PS	PS	PS	ZE
PS	NS	ZE	PS	ZE	NS
PB	NS	NS	NS	NS	NS

Table 14.2 Decision table for the fuzzy control of momentum constant

CCE	CE				
	NB	NS	ZE	PS	PB
NB	NS	NS	ZE	ZE	ZE
NS	NS	ZE	ZE	ZE	ZE
ZE	ZE	PS	PS	PS	ZE
PS	ZE	ZE	ZE	ZE	NS
PB	ZE	ZE	ZE	NS	NS

Reference [3] further improved the method and also presented some digital simulation. The results show that the convergence of fuzzy back-propagation learning is dramatically faster, and the steady-state value of the mean-squared error is significantly smaller than standard back-propagation learning.

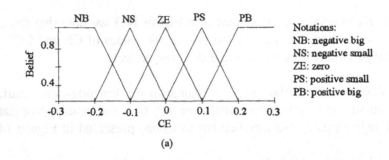

(a)

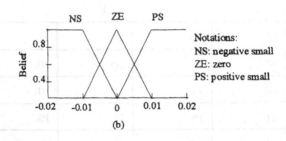

(b)

Figure 14.3 (a) membership functions for CE; the same membership functions are used for CCE; (b) membership functions for $\Delta\eta$

(i) Fuzzy neural networks

Figure 14.4 illustrates a fuzzy-centric hybrid system, which carries out fuzzy inference with NN structure, and adjusts the fuzzy parameters using NN learning. As discussed, fuzzy logic can be used to easily translate heuristic reasoning and knowledge from qualitative descriptions to quantitative descriptions. It has two important components: (1) fuzzy rules and (2) membership functions. Although fuzzy logic can easily capture the salient features of a given process, extract rules and membership functions are usually difficult to obtain. In this respect, neural networks can be trained to extract the rules and learn membership functions from some domain data. For example, as shown in Figure 14.5, multilayer neural networks cluster the linguistic variable inputs into membership functions for fuzzy space.

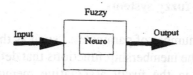

Figure 14.4 A fuzzy neural network

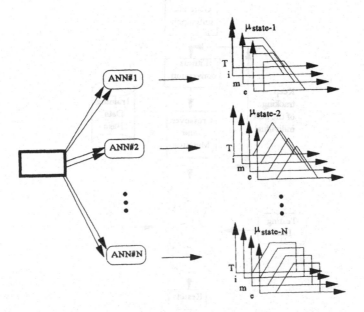

Figure 14.5 Learning membership functions by multilayer neural networks

14.2.2 Genetic neural networks

In the previous section, fuzzy logic has been employed to improve the training process and the NN accuracy. In this section, another attractive alternative - genetic algorithm - will be used to provide a good set of initial weights for the NN, or will be used to fully train the NN. The framework of genetic-based neural network is shown in Figure 14.6. After the initial populations are generated, an individual with high fitness (small output error of neural network) will be selected to mate for evolution. The best individual will be kept in each generation. The genes of this individual will be decoded to network weights. The gradient descent method is then applied for the back propagation in the neural network learning.

14.2.3 Genetic based fuzzy systems

A fuzzy system has a number of parameters, such as the fuzzy set used for input and output variables, the membership functions that define the fuzzy sets, and the structure and entries in the fuzzy associative memory (FAM). Some fuzzy systems also include parameters that assign a weight to each fuzzy rule, or FAM matrix entry, indicating its relative importance in the overall system output. All

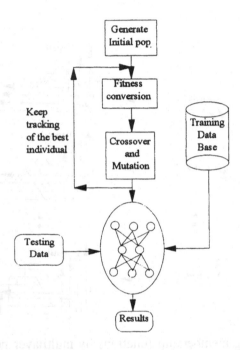

Figure 14.6 The framework of genetic-based neural network

of these parameters are candidates for adaptation with an optimization algorithm. While it is possible to design a system that adapts all of these parameters, this is a daunting task for the code developer. Common sense can provide pretty good estimates for the fuzzy sets and membership functions. The FAM matrix entries have the most influence in determining system output, so we will focus on adapting these.

Adaptation is accomplished through the minimization of an error function. However, in a fuzzy system, the parameters do not appear in an analytical way in the error function expression. Thus derivative-based approaches are not among the tools that we can bring to this problem. In this respect, genetic optimization can be effectively employed since it does not require analytic dependency of the error function on the parameters.

To apply genetic optimization to FAM matrix adaptation, we string the matrix entries together into a single long vector. Binary representation is used, so that each matrix itself is a vector of 1s and 0s. The result is a very binary vector formed by linking together the FAM matrix-entry binary vectors end to end. This is the chromosome upon which the genetic algorithm operates.

14.2.4 Fuzzy controlled genetic algorithm

As discussed in Chapter 12, genetic algorithms are distinguished by the emphasis on crossover and mutation, therefore more recently much attention and effort has been devoted to improve them. In this respect, two-point, multi-point and uniform crossover, and variable mutation rate have been recently proposed. More advanced genetic operators have been presented in ref[7] which are based on fuzzy logic with the ability to adaptively/dynamically adjust the crossover and mutation during the evolution process. Figure 14.7 presents the block diagram of a fuzzy-controlled genetic algorithm, in which two on-line fuzzy logic controllers are used to adapt the crossover and mutation. The objective here is to provide a significant improvement in the rate of convergence.

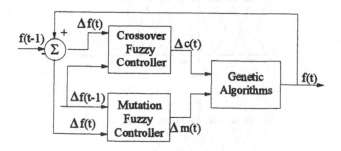

Figure 14.7 Block diagram of the proposed fuzzy controlled genetic algorithm

(i) Fuzzy Crossover Controller

The fuzzy crossover controller is implemented to automatically adjust the crossover probability during the optimisation process. The heuristic updating principals of the crossover probability is if the change in average fitness of the populations is greater than zero and keeps the same sign in consecutive generations, then increase the crossover probability, otherwise the crossover probability should be decreased.

(1) Inputs and output of crossover fuzzy controller
The inputs to the crossover fuzzy logic controller are changes in fitness at consecutive two steps, i.e., $\Delta f(t-1)$, $\Delta f(t)$, and the output of which is change in crossover $\Delta c(t)$.

(2) Membership functions of $\Delta f(t-1)$, $\Delta f(t)$, and $\Delta c(t)$
Membership functions of fuzzy input and output linguistic variables are illustrated in Figure 14.8. $\Delta f(t-1)$, $\Delta f(t)$ are respectively standardised into the range of [-1.0,1.0], and $\Delta c(t)$ is standardised into the range of [-0.1,0.1] according to their corresponding maximum values.

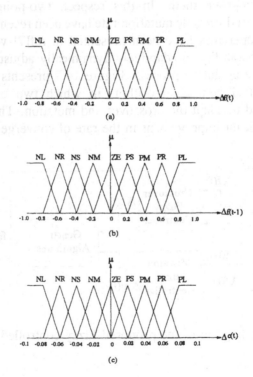

Figure 14.8 Membership functions of $\Delta f(t-1)$, $\Delta f(t)$, and $\Delta c(t)$

Where NL - Negative Large, NR - Negative Larger, NS - Negative Small, NM - Negative Medium, ZE -Zero, PS - Positive Small, PM - Positive Medium, PR - Positive Larger, PL - Positive Large.

Table 14.3 Fuzzy decision table for crossover

$\Delta c(t)$ $\Delta f(t)$ $\Delta f(t-1)$	NL	NR	NM	NS	ZE	PS	PM	PR	PL
NL	NL	NR	NR	NM	NM	NS	NS	ZE	ZE
NR	NR	NR	NM	NM	NS	NS	ZE	ZE	PS
NM	NR	NM	NM	NS	NS	ZE	ZE	PS	PS
NS	NM	NM	NS	NS	ZE	ZE	PS	PS	PM
ZE	NM	NS	NS	ZE	ZE	PS	PS	PM	PM
PS	NS	NS	ZE	ZE	PS	PS	PM	PM	PR
PM	NS	ZE	ZE	PS	PS	PM	PM	PR	PR
PR	ZE	ZE	PS	PS	PM	PM	PR	PR	PL
PL	ZE	PS	PS	PM	PM	PR	PR	PL	PL

Table 14.4 Look-up table for control action of crossover

y \ z x	-4	-3	-2	-1	0	1	2	3	4
-4	-4	-3	-3	-2	-2	-1	-1	-0	-0
-3	-3	-3	-2	-2	-1	-1	-0	+0	1
-2	-3	-2	-2	-1	-1	-0	+0	1	1
-1	-2	-2	-1	-1	-0	+0	1	1	2
0	-2	-1	-1	-0	+0	1	1	2	2
1	-1	-1	-0	+0	1	1	2	2	3
2	-1	-0	+0	1	1	2	2	3	3
3	-0	+0	1	1	2	2	3	3	4
4	+0	1	1	2	2	3	3	4	4

(3) Fuzzy decision table

Based on a number of experiments and domain expert opinions, the fuzzy decision table was drawn in Table 14.3.

(4) Look up table for control actions

For simplicity, a look-up table for actions of the crossover fuzzy logic controller was set up. Firstly, the quantified levels corresponding to the linguistic values of input and output fuzzy variables of the crossover fuzzy logic controller are designated, which are -4, -3, -2, -1, 0, 1, 2, 3, 4, respectively. Let x labels the quantified levels of $\Delta f(t-1)$, y labels the quantified levels of $\Delta f(t-1)$, and z labels the quantified levels of $\Delta c(t)$. Then the look-up table is formulated as Table 14.4.

In Table 14.4, $z = <\alpha x + (1-\alpha)y>$, where z means a minimum integer which is not greater than $\alpha x + (1-\alpha)y$. α is an adaptive coefficient which varies as the changes in the fitness of whole populations. It is found that good performances of the crossover fuzzy controller have been achieved when α equals 0.5. The output of the crossover fuzzy logic controller is formulated in equation (1),

$$\Delta c(t) = \text{look-up table}[i][j]*0.02*\beta \qquad (14.1)$$

where i, $j \in \{0,1,2,3,4,5,6,7,8\}$, the contents of the look-up table[i][j] are the values of z in Table 14.4, β is another adaptive coefficient which is less than 1.0 when the changes in fitness of whole populations are less than 0.02. Therefore the crossover is computed by equation (2).

$$\text{crossover}(t) = \text{crossover}(t-1) + \Delta c(t) \qquad (14.2)$$

(ii) Fuzzy Mutation Controller

The mutation operation is determined by the flip function with mutation probability rate, and the mutate bit is randomly performed. The mutation probability rate is automatically modified during the optimisation process based on a fuzzy logic controller. The heuristic information for adjusting the mutation probability rate is if the change in average fitness is very small in consecutive generations, then increase the mutation probability rate until the average fitness begins to increase in consecutive generations. If the average fitness decreases, the mutation probability rate should be decreased.

(1) Inputs and output of mutation fuzzy controller
The inputs to the mutation fuzzy controller are the same as those of the crossover fuzzy controller, and the output of which is the change in mutation $\Delta m(t)$.

(2) Membership functions of fuzzy linguistic variables of mutation fuzzy controller
The membership function distributions of $\Delta f(t-1)$, $\Delta f(t)$, and $\Delta m(t)$ are shown in Fig. 14.8.

Where the labels of the fuzzy subsets of changes in mutation are similar to those of the crossover fuzzy controller, but their meanings are different, as shown in Fig. 14.8 (c).

(3) Output of mutation fuzzy controller

Likewise, the fuzzy decision table and look-up table of mutation fuzzy controller are similar to those of crossover fuzzy controller, the control action of which is formulated by equation (14.3).

$$\Delta m(t) = \text{look-up table}[i][j]*0.002*\beta \hspace{3cm} (14.3)$$

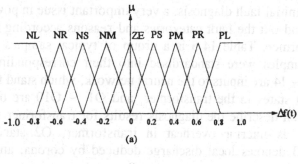

(a)

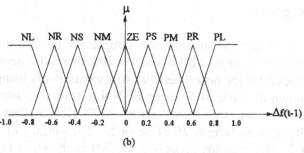

(b)

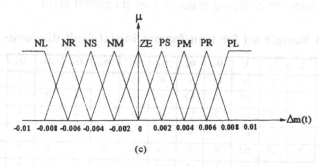

(c)

Figure 14.8 Membership functions of $\Delta f(t-1)$, $\Delta f(t)$, and $\Delta m(t)$

14.3 Applications in Power Systems

In recent years, hybrid intelligent systems have been applied to various power system problems such as: (1) load forecasting using fuzzy neural network; (2) static security assessment using genetic based neural networks; (3) power system stability using fuzzy neuro control. Here, two example applications are given to demonstrate the improved performance.

14.3.1 Fuzzy controlled neural networks for power transformer fault diagnosis

Transformer initial fault diagnosis, a very important issue in power engineering, attempts to find out the fault categories and reasons according to the gas states in the transformer. Table 14.5 is a group of typical samples taken from [6], where the samples were standardised into their corresponding discourses of universe. I1 ~ I4 are inputs to the neural network, which stand for the analytical results of gas states in the transformer, and O1 ~ O10 are outputs of neural network, which describe the diagnosis results for transformer respectively; for example, O1 is interior overheat in transformer, O2 stands for creeping discharge, O3 denotes local discharge deduced by corona, and O4 stands for slight discharge etc..

Using BP neural network to train the sample set for transformer initial fault diagnosis, the global error of BP is about 0.795 after 1400 iterations. Obviously, it is very difficult for the neural network to converge only using BP. The outputs of BP are shown in Table 14.6. Using FCNN to train the same sample set under the same constraints, the global error of FCNN reaches 0.0199 after 67 iterations. Its training time is 26.31 seconds. The outputs of FCNN are shown in Table 14.7. The training process of FCNN is shown in Figure 14.9. Figure 14.10 indicates the changing trajectory of its global error.

Table 14.5 Sample set for transformer initial fault diagnosis

No	I_1	I_2	I_3	I_4	O_1	O_2	O_3	O_4	O_5	O_6	O_7	O_8	O_9	O_{10}
1	0.5	0	0	0	1.0	0	0	0	0	0	0	0	0	0
2	0.5	0	0	0.5	0	1.0	0	0	0	0	0	0	0	0
3	0.5	0	0	1.0	0	1.0	0	0	0	0	0	0	0	0
4	0	0	0.5	0.5	0	0	1.0	0	0	0	0	0	0	0
5	0	0	1.0	0.5	0	0	1.0	0	0	0	0	0	0	0
6	0	0	0.5	1.0	0	0	1.0	0	0	0	0	0	0	0
7	0	0	1.0	1.0	0	0	1.0	0	0	0	0	0	0	0
8	0	0	0	0.5	0	0	0	1.0	0	0	0	0	0	0
9	1.0	0	0	0	0	0	0	0	1.0	0	0	0	0	0
10	1.0	1.0	0	0	0	0	0	0	0	1.0	0	0	0	0
11	0	1.0	0	0	0	1.0	0	0	0	0	1.0	0	0	0
12	0	1.0	0.5	0	0	0	0	0	0	0	0	1.0	0	0
13	1.0	0	0.5	0	0	1.0	0	0	0	0	0	0	1.0	0
14	1.0	0	1.0	0	0	1.0	0	0	0	0	0	0	0	1.0